Ehrenfried Pfeiffer

SENSITIVE CRYSTALLIZATION PROCESSES

A demonstration of formative forces in the blood

With 4 drawings in the text
and 26 plates containing 78 photographic
reproductions

Anthroposophic Press

Authorized translation from the German text
by Henry B. Monges
1936

Revised by Erica Sabarth
and Henry N. Williams, M.D.

Second Printing 1975

Library of Congress Catalog Card Number: 68-31125

Table of Contents

Dedication

This reprinting of the revised edition of Dr. Ehrenfried Pfeiffer's *Sensitive Crystallization Method,* first published in 1936, makes it possible to include a dedication to his memory and the continuing purposes of his work.

Throughout his life Dr. Pfeiffer's interest lay in searching for the forces active in the balance of the different realms of nature. Even during his early school years, he especially wanted to learn more about the finer forces, the underlying causes and the laws behind nature. As a student he first studied electrical engineering, and later, at the University of Basle, chemistry, botany and physics, finally obtaining his Master's degree in chemistry there. After finishing his university studies, Dr. Pfeiffer was able to devote himself completely to intensive research on the natural forces that had occupied him since his youth. His close relationship with Dr. Rudolf Steiner and anthroposophy were an essential help to him, especially in his later practical research when it became his goal to make visible those natural formative forces that had not previously been accessible through physical and chemical means.

In 1939, Dr. Pfeiffer was awarded an honorary M.D. degree from Hahnemann Medical College in Philadelphia in recognition of his research on the early diagnosis of cancer using the Sensitive Crystallization Method that he had developed and that is the subject of this book. Since then, the results of this research have been confirmed by numerous workers in this field (see Bibliography for specific references).

The deep insight into the laws of nature that Dr. Pfeiffer developed as a scientist also made it possible for him to perfect methods of research in other fields, notably agriculture. Here again, as in all his work, he emphasized the importance of synthesis in addition to the customary analytical methods. By this he meant seeing the whole picture rather than only its parts, that is, considering together all possible phenomena so that from their entirety a true final solution would arise. This approach led also to the development of Dr. Pfeiffer's specific chromatographic technique and method of interpretation. His continuing interest in agriculture resulted in a widening of his activities as an agricultural consultant and lecturer in Europe and the Middle East from 1930 to 1939, and in the United States from 1933 until his death in 1961.

Dr. Pfeiffer was not only a scientist of integrity, but also a loyal friend and teacher who was always ready to help the many individuals who came to him for advice. In his last years he renewed his efforts to help young people to find the right way to the tasks that awaited them. He strove to help them to find a deeper understanding of nature, to understand the importance of restoring and keeping the balance of nature, and, by rebalancing it where it is disturbed, to bring healing forces to bear.

In spite of his vast scientific knowledge and great wisdom, Dr. Pfeiffer's refusal to express opinions that were not founded on personal experimental experience was both a lesson and a goal for those about him. His path still stands as an ideal to be striven for in objective observation and the intuitive, yet scientific, approach to practical problems. His untiring efforts over the years to make visible the subtle formative forces behind and underlying the material manifestations of life and growth has helped many to recognize and establish the reality of these forces, which, in their deeper significance, were revealed and explained for our time by Dr. Pfeiffer's teacher, Rudolf Steiner.

We wish to express our appreciation for the photograph of Dr. Pfeiffer provided by Mrs. Adelheid Pfeiffer, to those whose generous contributions have helped to make this new edition possible, and to Dr. Gilbert Church and the Anthroposophic Press for their interest and for undertaking this reprinting.

Erica Sabarth
Henry N. Williams, M.D.

May 29, 1974
Spring Valley, N. Y.

Introduction

This book will presumably cause at first a doubtful shaking of heads among some of its readers, by others it will perhaps bè laid aside without due consideration because of its "quite impossible" statements. Of what do these "quite impossible" statements consist?

Dr. Pfeiffer says that by means of his method of crystallization—without taking into consideration the plant or the human being under investigation —far-reaching conclusions can be drawn about the nature of their constitution. The researching of human blood according to his method permits a judgment of the bodily condition of the owner of the blood and to a certain degree also of his mental state, especially in recognizing conditions of congestion, inflammation, of tuberculosis, sclerosis, cancer and many other diseases. It not only permits the researcher to "read" in the crystallization plate of the existence of diseases, but also to find the location of the disease in the body. He further states that his method permits the testing of remedies—whether they should or should not be employed—before they are tried on the patient.

That sounds indeed unbelievable at first glance. Especially the statement—to all appearances fantastic—that the crystallization picture enables the determination of the *locality* of the diseases. It compels us to suppose that the blood—indeed every drop of blood—pars pro toto (as the part so the whole)—reflects not only the entire bodily condition and all its essential pathological changes, but also that it contains formative forces which direct the crystallizations in such a way that the plate offers in a certain sense a topographical image of the body in which we can behold as, in a magic mirror, everything of importance.

We medical doctors have had to recognize a great deal as justifiable—especially in the modern age—which formerly we ridiculed as senseless. We have therefore by rejecting "outsider-methods" become somewhat cautious and prefer to test first and judge afterwards. Frankly speaking, the Pfeiffer indications go beyond all our previous concepts, they require a complete change in our thought processes and hence it is entirely comprehensible that it awakens great mistrust and that but few have the desire to undertake the extremely painstaking and time consuming work of verification.

Is this testing of the facts worth while; can we hope for a positive gain for diagnosis and therapy with the Pfeiffer method? Let us concentrate on the illustrative material which Pfeiffer lays before us here! Even the most skeptical will have to grant that forms (pictures) which are produced by the plant extract crystallizations, are very remarkable. What Pfeiffer tells us seems quite believable, that every plant extract always produces only its own specific form type. Figures No. 21, 22, 23 and 24—with their interpretation by Pfeiffer together with the additional report of the contributor of the experimental material itself which is in agreement with it—offer much for our careful consideration and convey favorable impressions of Pfeiffer's statements. The beginner, however much already inclined to believe in the practicability of the method, is again taken aback when he observes the blood crystallization, reads the text connected with it, and is unable to find at all "clearly" the forms—of which Pfeiffer speaks—in many of the crystallization plates. Indeed he will be surprised that with such unclear material Pfeiffer's diagnosis can agree so exactly with the diagnosis of the hospital intern. I was doubtful at the beginning whether on the basis of *this* crystallization material I should set about verifying it, because the blood crystallizations, which reflect many possible bodily conditions, are grantedly much more difficult to read than are the simple plant crystallizations and they present an apparently almost unachievable task to the beginner. I was acquainted at the commencement of my research nine months ago with the plant crystallizations only and with the few—not very clear to me—blood crystallizations which Pfeiffer had published in his work, "A Study of the Formative Forces in Crystallizations". The plant crystallizations however awakened in me a hope that the blood research about which Pfeiffer at that time had reported only little might also be interesting and compensating.

Through the kindness of Dr. Pfeiffer I became acquainted with S. Rascher, M.D.—mentioned a number of times in this book—who proved to be an excellent teacher of the very difficult and delicate technique of the method and an untiring co-worker.

On the basis of my own experience, I can now say that weeks are necessary to be able to produce crystallizations which are of any use, and months are necessary in order to read them properly. This method, as Pfeiffer himself also says, will not become in any sense of the word a *universal* method of disease diagnosis and cure. Anyone who undertakes this method without possessing the necessary patience and technical ability, who is not gifted with a sense of form perception and intuition, should not attempt it, for he will only have failures which will then be charged against the method. To anyone however who takes pleasure in verification, I can say with encouragement, that the crystallization plates are much clearer and more easily read than the photographs of them; even when these are well done they are never able to reproduce all the fine details and important signs of the plates themselves, such as for example the variations in plane and smoothness of the crystallizations, likewise the color (variations) of the individual parts.

From our own results I am able to inform those who are interested that our own diagnoses very often in all their details, also in regard to localization, agree essentially with the hospital diagnosis. We have never had a complete failure in our diagnosis of the blood crystallizations. The patient from whom the blood was taken was sometimes known to Dr. Rascher—but intentionally he was always unknown to me. The crystallizations were diagnosed by both of us—first in complete secrecy—then the results were compared and reported in writing to the hospital. The description of our diagnosis and the hospital diagnosis were then entered in our record book.

In spite of our results, I should still not consider it proper for me at present to stand for all of Dr. Pfeiffer's proposals. Because of the uncertainty also of many hospital diagnoses I shall and can do this only after I have been convinced by numerous cases—as a result of the subsequent operations or dissections—of the real reliability of the Pfeiffer method. So far this has not been the case.* Yet I have now the impression already that it is really worth the trouble to occupy myself diligently with these facts, therefore I do not hesitate to invite my medical confreres to co-operate.
Munich, 12. July 1935.

PROF. DR. TRUMPP

---

* Footnote of the translator:

   Since the above was written this verification has been made and published by Prof. J. Trumpp M.D. and S. Rascher M.D., in "Münchener Medizinische Wochenschrift" 26. June 1936 under the title: "Nachprüfung der E. Pfeiffer'schen Angaben über die Möglichkeit einer kristallographischen Diagnostik; Versuch einer Hormonoskopie und Schwangerschaftsdiagnose.

Preface

The problem of "formative forces" is treated in the author's two books "Crystals" and "Studies of Formative Forces in Crystallizations".* In these publications an attempt is made to show that the processes of crystallization under certain experimental conditions are "sensitive"** to influences which produce an alteration of form. These processes of crystal formation can be greatly influenced—in regard to the subsequent arrangement of their characteristic crystal shapes—by the addition to the solutions, of small quantities of foreign substances—chiefly those derived from living organisms. But chemical telekinesis may also be observed.

It was clearly established that the change of form within relatively wide limits is quite independent of the chemical and physical nature of the admixture. On the other hand the greatest influence comes from the "organic" nature (quality) of the object under test, from which the admixture is taken. Its quality and vitality, state of health or disease and evolutionary history arrange themselves either in an *orderly or chaotic* manner. The reaction upon the form-producing forces, which the crystallizing salt solution shows as a "detector" or "reagent", justifies an *a posteriori* conclusion concerning the general nature of the agent used.

The sensitive crystallization-reaction becomes a research method by the help of which the vitality, the biological activity of the organism may be investigated. For example—of two similar seeds, the one derived from the healthier plant with the strongest germinating force, produces the more harmonious formation.

Thousands of crystallizations have been made over a period of years in the Chemical-biological Section of the Research Laboratories at the Goetheanum—with the assistance of and in connection with the Natural Science Section of the School for Spiritual Science (The Goetheanum), Dornach, Switzerland. This work began and was carried on during the years 1922–1923. Suggestive indications from Doctor Rudolf Steiner gave to our

---

* Published by E. Weise's Buchhandlung, Dresden, Germany, 1930 and 1931 respectively.
** When we employ the expression—"sensitive"—the psychological meaning of the term is of course not meant, but it is used in the same sense as when in exact Natural Science we speak of a "sensitive" seismograph or of a sensitive indicator, such as methylorange and so forth.

work the direction which we since have followed and in which our researches are now being made. He foresaw the possibilities of applying this method, but without his having had the opportunity during his lifetime of seeing the later practical results which up to that time had not been attained. My gratitude belongs primarily to him for the suggestive stimulus he gave me in my search for a "method of determining reactions due to formative forces" and the "establishment of the difference between various animal and human organisms through the quality and nature of their blood".

In both publications above mentioned, facts were given about the influence of plants, of food stuffs and of nature products in general on crystal formations. The formative-force values—specifically characteristic of every organism—were observed and demonstrated in harmony with general natural law. As soon as the first technical difficulties of this method of research were overcome, its extension to the experimental investigations of the blood of healthy and diseased human beings followed as a matter of course. The results were surprising and of novel and unusual character. Nevertheless I believe that this research at its present stage warrants the publication of its findings. A bio-chemical laboratory like ours was not able to carry on such work alone. It required the co-operation and support of physicians. After the technical aspect was solved a large number of physicians lent their aid; they came to our laboratory and assisted in working out the "method of sensitive crystallization" (crystallization response or reaction) for the experimental investigation of human blood. They helped greatly by giving "cases", by checking the results, by advisory council concerning the determining of specific characteristics of the individual aspects of disease symptoms.

The following pages represent the result of the collaboration of many people. They contain only an "extract" as it were of the results of many hundreds of crystallizations. It is not possible however to publish all the material at hand, even were the publishers willing to risk the high cost of such an undertaking.*

---

* We are constantly reproached with not having published a sufficient number of parallel results of comparative experimental research in spite of the fact that we have only followed a widespread scientific practice, hence we present here as a proof of this the conditions for the publication of scientific research results as laid down by the "Magazine for the Entire Field of Experimental Medicine" (Zeitschrift für die gesamte Experimentelle Medizin). Here we read among other things: "It is the rule to offer from each kind of experiment, respectively from each of the facts, only one report (history of the patient, dissection report, experiment) as an example, in telegraphic style, in the most concise form possible. The rest of the proof matter may be employed in the text or in tabular form when this cannot be avoided. It is advisable to point by means of a footnote to the institute where the entire proof material can be seen or borrowed. For each sort of experiment only one series of illustrations is permissible."

The technical difficulties of making the photographic reproductions were almost insuperable. We had to limit ourselves to the choice of some typical examples. The possibilities only of the method will be shown here. Investigators or researchers who wish to pursue the matter in more detail will have to spend some months working in our institution before they are able, even to some degree, to survey the whole field. With the appearance of new methods of solving a problem, the demand is often made for a "universal cure"; "there must be a cure-all!" The present method makes no such claim. It replaces neither the diagnosis of the physician, nor supplies the evidence for discovering a "universal remedy"; but it does enhance the possibility of knowledge and observation of everyone who earnestly studies the subject. It is in this sense that we wish this modest publication to be accepted.

Our gratitude extends to a large number of coworkers who have mastered, through many patient years of work, the technique of making the crystallizations: Erika Sabarth; Baroness Mia von Mackay, M.D.; Sigmund Rascher, M.D.; Johannes Teichert, Chemist; Helene Egger, Chemist; Marie Lommel; G. v. Volkmann and many others.

The author would like to mention especially, with thanks, the physicians through whose co-operation only was it possible to carry out the medical part of the work—Ilse Knauer, M.D. Berlin and Jean Schoch, M.D. Strassburg. Other physicians were A. G. Degenaar, M.D.; Richard Schubert, M.D. Dornach; Werner Kaelin, M.D. Arlesheim; E. Weil, M.D. Strassburg; Josef Kalkhoff, M.D. Freiburg i. Br.; likewise others who have given full measure of co-operation.

Worthy of special mention is the excellent and most painstaking photographic help of Mr. Emil Gmelin, Dornach.

The author owes a large measure of thanks for the helpful advice and generous active help which Professor Dr. Trumpp of the University of Munich has given in the testing and further pursuit of this work.

The practical value of our experiments lies in the fact that we have already been able to help several hundred people by means of the "blood crystallization" diagnosis. Not every physician however will be able to establish a crystallization laboratory of his own. The sensitiveness of the method makes especially high demands on the time and devotion of the researcher and requires special equipment which is in all cases very expensive.

The guiding idea underlying the use of the terminology of this book was to write it in a way that could be understood by everyone; to this end we have avoided as far as possible all technical expressions. We are aware

10

however that this work is just the first beginnings of an entirely new sort of research. It will therefore undoubtedly call forth many criticisms. We recognize also that it will have to be greatly improved—but the author is convinced that whatever the present shortcomings, the consequences of coming in contact with the nature of the "formative forces", or the "etheric formative forces" (in nature) can contribute much to the deepening of the knowledge of the individual who sincerely seeks to investigate this heretofore little known field. If through the working out of this method something can be contributed to the health and healing of diseased human beings then the hope which accompanies this book will be fulfilled.

Dornach-Switzerland, July 1935.

EHRENFRIED PFEIFFER.

The translator wishes to express his thanks for valuable assistance rendered by Miss Saunders-Davis and Frl. Lisa Dreher.

Preface to Revised Edition

Refinements in technique over the past thirty-one years have required a complete revision of the last chapter on "Technical Details". Rising costs have dictated reducing the plates of photographic reproductions where duplications existed in the original publication. Comprehensiveness requires the insertion of a bibliography to which the reader is referred. Personal gratitude must be expressed to the author for his assistance (both personal and by means of the Sensitive Blood Crystallization Test) in my practice and research, and to his coworker Erica Sabarth (since his death in 1961). I also wish to express my appreciation to Mr. Richard E. Huss, printer and bookbinder, for his willingness to take on the technical problems of this revised edition, and to Mrs. Adelheid Pfeiffer and Mrs. Henry B. Monges for allowing this revision to be published.

Henry N. Williams, M.D.

Lancaster, Pa. January 1967

SENSITIVE
CRYSTALLIZATION
PROCESSES

I. A Method of Producing "Sensitive" Crystallizations

Research carried on since 1925 has shown that the formation and arrangement of crystals during the process of crystallization can, under certain conditions, be greatly influenced by the admixture of various substances. The alteration in the crystal formations is determined by means of the nature of the admixture. Hence, from these alterations (in form) a priori conclusions can be drawn about the qualities and characteristics of the admixture itself. A large amount of illustrative material has already been introduced in previous publications of the author.*

The technique is essentially as follows:—Let 10 cc. of a 5% to 20% solution of copper chloride ($CuCl_2$) be poured out evenly over the surface of a circular glass plate 10 cm. in diameter. As a result, the glass plate is covered with a solution of not more than ½ to 1 mm. deep. This is then placed on a vibration-free table in a chamber, the temperature of which ranges around 28 degrees centigrade. The relative moisture content of the air should not be below 30% nor above 50% of the chamber's cubic contents. Because of the gradual evaporation of the solvent (menstruum)—water—

---

* E. Pfeiffer, "Kristalle" (Crystals), with 94 illustrations, published by E. Weises Buchhandlung, Dresden, Germany.—E. Pfeiffer, "Studium von Formkräften an Kristallisationen" (A Study of the Formative Forces in Crystallizations), with 55 illustrations, published by the Science Section of the Goetheanum, Dornach, Switzerland. A condensed edition of this work is published by Rudolf Steiner Book Center and Publishing Co., London, under the title: "Formative Forces in Crystallizations."

crystallization occurs. If this takes place between the 14th and 18th hours after pouring, we obtain a result, the purpose of which will be described later on.**

If an admixture of from 1 to 3 drops of a plant extract, of various (by no means all) inorganic salts or dilute solutions of animal or human substances, such as saliva, urine, the extracts of bodily tissues and finally of very much diluted human blood, be added to the salt solution prior to its being poured out on the glass plate, the normal crystallization will be essentially altered in general in its form structure.

This phenomenon may be compared with the delicate flower-like ice structures which appear on window panes in winter as the result of the freezing of water vapor.

As an introduction, we shall begin by repeating some of the examples and illustrations from previous publications.

Plate I

Fig. 1:—This shows the crystallization of a normal 20% solution of chloride of copper under the stated conditions. In all the so-called "controls"—i. e. crystallizations without admixture, it is characteristic that crystals are rarely deposited on the glass right out to the circumference; but compress themselves more toward the centre, forming a confused muddle. With the greatest care and exactness in the experimental conditions, we obtain practically the same typical control characteristics in crystallizations which have been made daily now over a period of ten years.

Fig. 2:—A 25% solution of copper chloride with an admixture of 1 drop of fresh extract, made from the sap of the water-lily. Attention is here called to the resultant plantlike arrangement of the copper chloride crystals.

Fig. 3:—A 25% solution of copper chloride with an admixture of 1 drop of the juice of the leaf of the American Aloe (Agava Americana). You will note here the prickly, radiating character of the resultant chloride of copper crystals.

Figures 2 and 3 are exceptions which are made with alcoholic solution, all the other crystallization plates are made with water solution.

Fig. 4:—This is the same copper chloride solution with an admixture of 1 drop of chamomile blossom juice. These experiments show that with each plant juice or sap there results a crystal form or structure typical of it and of it alone. Repetitions with the same sort of admixture produce exactly the same typical form in 80% of the cases.

** Concerning the technique, see exact details in chapter V.

Plate II

Figures 5–8:–Show simultaneous crystallizations of a 5% copper chloride solution with an admixture of honey (5 drops). The result is the same typical form.[*]

These four illustrations are taken from a series of 9 exactly similar pictures.

The chemist and physicist will make the objection that the observed phenomenon is produced by the change in the surface tension, the colloidal state of the admixture or by chemical reaction and so forth. The following series shows the difficulties of such an explanation.

Plate III

Fig. 9:–Copper chloride 20% solution without admixture.

Fig. 10:–Magnesium sulphate 20% solution without admixture.

Fig. 11:–Acetate of lead 10% solution without admixture.

All these salts are chemically and physically different. Each of these salt solutions now receives the same admixture. This consists of 3 drops of freshly drawn human blood, diluted in 1 cc. distilled water; of this solution 3 drops are admixed with each of the three salt solutions.

Fig. 12:–Copper chloride 20% solution, with admixture of diluted human blood.

Fig. 13:–Magnesium sulphate 20% solution, with admixture of the same as above.

Fig. 14:–Lead acetate 10% solution with admixture of the same.

The same typical structure–a crystal radiating from a single centre to the circumference–appears in *all three different salts* containing the same admixture. The form producing influence of the admixture is thus able to assert itself in spite of the chemical and physical variations of the crystallizing salts. It is therefore permissible to speak of the effect of special formative forces which prevail over what happens directly in the physical and chemical elements and also which influence these occurrences.

These effects are to be especially observed in the greater (higher) dilutions of the admixture. However in a moderate concentration of the admix-

---

[*] In E. Pfeiffer's "Studies of Formative Forces in Crystallizations", illustrations are given, in plate I of a series of ten crystallization "controls" and in plate II of a series of ten crystallizations with the admixture of the sap of St.-John's-wort (Hypericum), as a further proof of the regularity of the phenomenon. For lack of space we are compelled to forego any further reproductions of the same sort of series here. Naturally it should be understood that in each experiment a number of crystallizations are made.

ture the chemical and substantial influences predominate. With increased dilution the latter influence is reduced and gives place to the formative influence of the admixture. The maximum (reactive) sensitiveness can be attained with a certain dilution which has only been ascertained empirically.

The reaction produced is thus the resultant of:—

(a) Temperature, atmospheric humidity, lack of vibration in the experimental cabinet (or chamber), absence of drafts or air currents, in general the nature and condition of the crystallization chamber or cabinet;

(b) degree of concentration (potency) of the solvent and of the admixture;

(c) duration of the crystallization period, further instances of the different, varying influences, to be discussed in chapter V.

The forms obtained are a resultant of the copper chloride's own formative force and of the form-producing influence of the admixture.

Plate IV

The following example of the influence of an inorganic admixture may serve to show the influence of the dilution (as such).

Fig. 15:—10 cc. of a 10% solution of copper chloride with an admixture of 100 drops of a 10% solution of bicarbonate of soda (natrium bicarbonate). The reaction produces an aggregate of crystals which shows, according to the degree of concentration, a more or less characteristic type.

Fig. 16:—The same with 50 drops of bicarbonate of soda solution.

Fig. 17:—The same with 30 drops of bicarbonate of soda solution shows struggle of the form-producing (formative) tendency of both salts.

Fig. 18:—The same with 10 drops of admixture. The sensitive resultant appears.

Fig. 19:—The same with 5 drops of admixture results in the most favorable (optimal) effect.

Not all inorganic salts however give such results. They have been observed with potassium chloride, natrium chloride, natrium carbonate and bicarbonate; potassium chloride, iron ammoniumsulphate, potassium permanginate, natrium acetate; further, cane sugar treated biologic-dynamically, lecithin and gelatin. On the other hand other salts and substances such as calcium chlorate, calcium nitrate, ammonium chloride, synthetic citric acid, tartaric acid, acetic acid, nitrate acid, soluble starch, beet sugar, ethyl- and methyl-alcohol, glycerine, formaldehyde and paraldehyde (Fig. 20) have not at all influenced the copper chloride. With the latter substances, crystallizations appeared just as they are known through their "controls". A comparison of the above-mentioned salts supports the conjecture that *the most sensitive reaction predominates in those* substances which play a fundamental physiological role in the up-building of living organisms. On the other hand, failure was experienced with substances of an inert laboratory origin as for example potassium chlorate, formaldehyde and paraldehyde.

The extent to which the sensitiveness of this method of crystallization can be carried can be shown in the following two examples:—For a number of years we have used this method to test the quality of agricultural products, and amongst other things, the germinatable and cultivatable quality of seed. The seeds were germinated in water and were then carefully crushed in a porcelain mortar, avoiding contact with all metal or the human hand. Two or three drops of the filtered extract were added to 10 cc. of a 5% copper chloride solution.

Plate V

Fig. 21. This picture was obtained by the addition of an extract from pine seed, sent to us by a forester, together with other specimens, with the request to submit a report. The washed out, unclean formation of the crystals—interrupted, encrusted and broken throughout—is very noticeable. The crystals do not radiate from one centre to the circumference. In short it is thoroughly malformed.

Fig. 22. The pine seed of another tree in the same district gave this remarkably delicate and finely organized structure. From one centre the crystals extend to the circumference in uniform symmetrical lines. The formation in this crystallization is the direct opposite of the foregoing.

Inasmuch as the seed of a plant epitomizes, as it were, the whole life history of the parent plant, i. e. its whole structure has been produced and influenced by the course of its growth, we can make certain inferences from the crystallization of this seed-extract regarding the parent plant. In this latter instance we have to do with a healthy, normally developed tree; in the former case with a stunted growth. This report was forwarded to the forester who sent the specimens, and by return post we received a photograph of the trees in question and were astonished at the accuracy of our conclusions.

Figures 23 and 24. Photographs of the two trees. Right, (Fig. 24) shows the well developed pine tree with its full crown, the seed of which gave Fig.

22. Left (Fig. 23) shows a crooked pine with a stunted crown, the seed of which gave Fig. 21. (The photograph is taken from the base of the trunk looking upward.)

Plate VI

Fig. 25. Another similar example. Specimens of beechnuts were received from Mecklenburg for examination. One group of seed gave a beautifully symmetrical crystallization. (The seeds were germinated as before, pressed, and a few drops of this were mixed with a 5% copper chloride solution.) The resultant crystals radiate from a single centre to the circumference.

Fig. 26. Another group of seed gave a disconnected crystallization. Many centres appear, radiating outward asymmetrically. The individual centres are characterized by glittering, winglike forms. A uniform structure is not discernible.

We may immediately conclude that the seeds which gave Fig. 25 originate from healthy trees, while those which gave Fig. 26 come from unsound stock. But the plate reveals still more. From earlier researches in connection with root-extracts it has been observed that these generally give forms analogous to those found in Fig. 26. We have now this curious fact, that in this plate it is seed, and not root which produce these forms. It follows that the poor development of the formative forces in the seed may be traced to an unhealthy condition of the root of the parent tree.

This report was sent to the forester of the Mecklenburg beech forest and by return mail came the solution of the enigma. The underground water level of the district had changed with the result that a part of the beech forest was standing with its roots in dried out land. The trees were beginning to sicken and die out. This process shows its effects in the constitution of the seed and in the crystallization of the seed-extract (Fig. 26). The seed which gave Fig. 25 did, as a matter of fact, come from sound trees.

II. Experiments with the Blood of the Healthy and the Diseased

After having shown in previous publications the application of this research method in the various fields of biology, foodstuffs and so forth, it was obvious that the next step would be an extension of its possibilities to a field, especially interesting to Doctors of Medicine, namely the field of research in human blood. During the years 1930–1935 intensive researches in this direction were undertaken after having observed the effects of the admixture of highly diluted hemolysed human blood on the crystallization of $CuCl_2$ which showed surprising differences of form, depending upon whether the blood originated from healthy or unhealthy persons, also upon the variation of disease. Our fundamental observations are set forth in the following pages.

A blood crystallization picture is called normal, when it is made with 10 cc. of a 20% solution of copper chloride, admixed with 3 drops of diluted blood; the degree of this blood dilution is described in chapter V. This solution should crystallize between the 14th and 18th hours after having been poured out upon the prepared glass plate. The crystallization process will be disturbed if the evaporation takes place either too rapidly or too slowly. Given this normal lapse of time, the crystals arrange themselves around a central point, radiating thence in all directions to the circumference of the glass plate. This point of radiation is rarely found in the exact centre of the plate. It is generally somewhat excentric i. e. it is displaced to one side of the real centre. We call this the "centre of gravity" and we hold the plate in such a way that when looking at it horizontally, the narrower division lies toward us (i. e. the space between the "centre of gravity " and the plate circumference lies toward the observer).

This normal crystallization with an admixture of human blood is rarely obtained. Such a plate is shown in Fig. 12. The crystallizations of different individuals may be extraordinarily diverse. The simplest variation is that in which the radiating crystalline formation does not extend to the extreme edge. The dense rays cease at a certain distance from the periphery, from about 1 mm. to 1 cm., and a fine, more or less distinct network of crystals forms a thin layer as a border; or there may appear encrusted areas or spaces, resembling somewhat the control plate, i. e. the crystallization of copper chloride made without admixture. This means that the formative force of the admixture has been reduced or is entirely absent.

A further variation is the appearance of many different centres. These may either be grouped around the "centre of gravity", or may be scattered irregularly over the whole surface; or again, they may appear only around several single points. For further diagnosis however in such a case, it is important to determine a conjectural, "theoretical" centre of gravity. Let us imagine a disk on which accumulations of matter are excentrically scattered. By purely geometric-mechanical means we can find this conjectural "centre of gravity" by which the stable position of the plate may be determined. An aesthetic sense can assist us in finding this point *instantly*, at a glance. We can also call this ideal point the centre of gravity, and use it in the same way for the purpose of orientation in the manner previously described when refering to the true centre.

The next observation shows different types of centres. Some are two-winged, some three or four-winged. Some have wings on a base line and others have radiating forms. From twenty to thirty and more different types may be defined. In their arrangement the crystal groups may radiate or be intercepted with large round, long oval or triangular spaces between them. These further may be distributed generally or clustered together. Further combinations among all these forms are possible. Even the coloring may vary, the normal blue-green color may at times lighten to a yellowish green as a result of accumulations of water; the water adheres to the crystals. Are all these differences to be attributed to mere chance or can a deeper cause be discovered? This was the question which the author asked himself. Having obtained regular formations in the crystallizations in the manner already mentioned with plant extracts, the next logical step was to look for a connection based on natural law.

The first thing we noted was that the so-called normal blood crystallization appeared in experiments made upon those who felt "well". By this we do not mean the condition of a robust athlete, but a feeling of bodily well-being, of a vitality, free from fatigue, pain or any abnormal condition. The blood of a person which produced a normal copper chloride crystallization at one time, a few months later, during an attack of fever, may produce an entirely altered crystallization.

We then made arrangements for a systematic co-operation with various doctors, which resulted in our receiving from them cases of well-defined diseases for investigation. In order to work as free from bias as possible, the blood-tests were made without any knowledge of the medical report on our part. The characteristics of the plate were first carefully noted and described by us and then this result was carefully compared with the medical report. Certain diseases were found to correspond to the presence of very characteristic deformations of the normal blood-crystallization. Definite forms appeared when the patient was suffering from a definite disease. There were cases in which crystallizations at first appeared to contradict one another or could not easily be explained. Here still finer distinctions had to be observed. To-day after five years of practical application of the method in this field we feel compelled to say that the smallest modifications have to be taken into account and that out of some 300 cases only five still remain obscure. We have made more than 30,000 crystallizations, among which there were several thousands with a blood admixture; besides this there were more than 400 verified cases of disease. This was enough material for comparison in support of the results obtained. Also in other localities similar researches have been already undertaken in collaboration with the author with confirmative results. The surprising result, the description of which follows, accounts for the long delay in publishing this brochure.

The arrangement of the crystals on the glass plate corresponds to definite laws. From a consideration of the position of the centres in their relation to the "centre of gravity", and to an imaginary cross formed by the vertical and horizontal axes intersecting at this centre of gravity, it is possible to locate the corresponding symptoms in the organism of the patient. This means that it is possible to determine the seat of the illness from the relative position of the typical form in the crystallization picture. (See diagrams 1 and 2.) Thus a deformation in the narrow area between the point of intersection of the two imaginary axes and the edge of the glass plate corresponds to affections of the head or the nervous system and the organs of sense. In the wider zone, in the immediate vicinity of the horizontal axis, the processes of the respiratory and circulatory systems are mirrored, while the regions still further removed upward towards the periphery, represent the processes of metabolism. An exact localization can thus be obtained. Care must of course be taken to avoid a too mechanical interpretation. Thus if an abnormal centre appears in a particular place on the crystallization plate it will also affect its neighbourhood and it may push some or all of the other centres out of their normal and habitual position. The ability to interpret these "displacements" in crystallizations can only be acquired after much experience and through the observation and comparison of numerous crystallizations. Only a person who is able to think about the possible modifications in a vital way and who possesses intuitive understanding, will be able to find his way among the manifold forms and be competent to distinguish the regular and typical form from the accidental and uncharacteristic form.

The following description of a few characteristic types can best illustrate what has already been said. First we shall describe the blood crystallization and single out its characteristics and then we shall draw the possible conclusions. We shall then show the doctor's report. In practice this was the actual procedure followed. In some instances the original report and corresponding medical report will be given here to show the method of collaboration. In order to facilitate description, we shall employ the following nomenclature which will become clear by reference to diagrams 1 and 2.

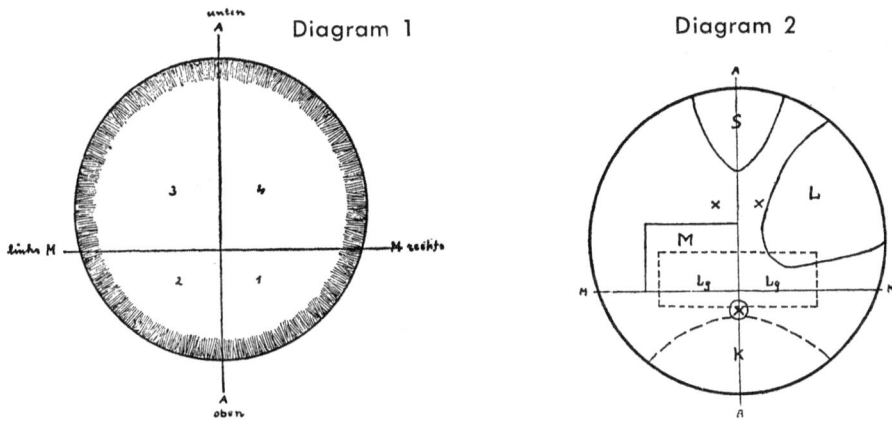

Diagram 1   Diagram 2

A-A = vertical axis; M-M = horizontal axis; the intersection of the A-A and M-M- axes = centre of gravity; numbers 1, 2, 3, 4 show four quadrants of the crystallization field; the hatched portion = peripheral zone; S = zone of the sexual organs; M = gastric zone; Lg = pulmonary zone; X = kidney points; ø = throat zone; K = head zone; L = liver zone.

Plate VII

Fig. 27. The first plate of this series shows rather finely formed crystallizations in the peripheral region. The border zone is ill-defined. Several crystal groups are arranged in such a way that the form of a cross appears, at one point distinctly, at other places imperfectly. The arms of the cross are narrowed towards the point of intersection and thence broaden outward roughly resembling the shape of a maltese cross. Often the space between the arms is a void, often overlayed by other groups of crystals or by the crossed arms of another crystal group. The most distinct form is to be found near the centre of gravity; the others somewhat below the horizontal axis are displaced to the right and to the left. In the third quadrant three-winged forms appear frequently, the fourth wing however is not developed. Owing to the photographic lighting these typical forms stand out somewhat more darkly than usual. Empty spaces are distributed over the whole blood crystallization, but these are of secondary importance. A more clearly defined four-winged form appears in the fourth quadrant, somewhat near the one first described.

Fig. 28 shows the same plate as Fig. 27, but differently photographed. Fig. 27 was photographed under an illumination from behind, while Fig. 28 was taken under direct lighting. The characteristic centres appear in one case quite different from the other. The plate should be examined under every possible lighting in order to distinguish carefully the finer details. Unfortunately with photographic reproductions it is not possible except to a limited degree to render the fine details and precision of the originals.

Fig. 29 shows an enlargement of the most important centre. The form of a cross in the crystals is very noticeable. The crystallization is that of a patient aged 36—female—(case No. 262). The medical diagnosis* certified by Dr. J. W., is as follows:

9th and 12th year	: Pleuritis sicca et serosa,
26th year, autumn	: Pleuritis sicca,
27th year	: Pleuritis serosa,
32nd—35th year	: Tuberculosis pulmonum et laryngis,
36th year	: Cavity in the right lung.

By a comparison with the diagnosis we obtain a clue to localization. The centre of gravity—the cruciform crystallization—corresponds to the region of the larynx. The clearly defined cross formation—see enlargement—is attached to the right and to the left, the right being somewhat more clearly defined than the left.

Plate VIII

Fig. 30. The next example—case 402, male, aged 30—again shows a distinct deformation in the neighbourhood of the centre of gravity, and in the inner corner of the fourth quadrant. Here again the crystals have the

---

* A resumé of the diagnosis will suffice for the present text. A report of each case is preserved in our files. Further details can be obtained from the attendant physician.

tendency to begin from a point, narrow at first, broadening out into wings. Three and four-winged forms appear. In this plate, however, there are no hollow spaces as in the previous case. By comparing the various cases it has been observed that the extent of the hollow spaces—number and size— denotes the gravity of the case. The preceding case, in comparison with the present one, reveals a more advanced stage of the disease. Points of disturbance lying farther toward the periphery in the third quadrant are not so characteristically formed as is the case with the central groups. We have used the central group as the basis of judgment; see Fig. 31.

Diagnosis by Dr. K. and others:—Pulmonary tuberculosis, the right lung more gravely affected than the left. The border zone of this plate is considerably more clearly defined than that of the preceding case. It bears testimony—according to our experience—to the vitality of the patient. In the preceding case the vitality is weaker. Here it is still fairly strong.

From the foregoing as well as from 19 other cases we perceive that there is an unequivocal connection between tuberculosis and the appearance of what has been designated simply as the "Maltese Cross" form. In the cases of pulmonary tuberculosis this form appears near the centre of gravity in the inner angles of the third and fourth quadrants (Lung-zone, diagram 2), in cases of renal tuberculosis, more toward the periphery in the same quadrant near the vertical axis (kidney-point, diagram 2), and directly on its centre of gravity in the cases of laryngeal tuberculosis. In cases of intestinal tuberculosis the Tbc. centres preponderate in the third and fourth quadrants, distributed in a horizontal line one third of the distance from the lower margin. By comparing such crystallizations with the diagnosis we are inclined to conclude that tuberculosis is present. But a still further proof is needed. The plates we have used so far were made with blood admixtures of various patients. Thanks to the excellent work of a medical student, Mr. Sigmund Rascher (now M.D.), working in our laboratory, it was possible to establish these proofs which are here briefly described.

Mr. Rascher received, during his student period in the medical school of the University of Basel, specimens of the diseased organs taken during operations and from dissections of tuberculosis patients. He made a filtered extract with water, and following our procedure he added several drops of this extract to 10 cc. of copper chloride. Crystallization occurred in the usual way and the results are the following plates.

Plate IX

Figures 32–35. Extract made from a pathological specimen obtained from a portion of a tubercular lung of a male, aged 42. Here we see four times repeated in the copper-chloride crystallization, the extraordinarily well-defined form of the "Maltese Cross" with the voids around the centres.

Plate X

Fig. 36. Specimen obtained during operation from a tuberculosis fungiformis cubiti dextrae, male aged 18.

Fig. 37. Specimen obtained during operation from a tubercular periostitis of the head of the ulna (capituli ulnae dextrae), female, aged 70.

Plate XI

Fig. 38 is the blood crystallization of a patient, aged 63 (case No. 410). Opposite the centre of gravity is a striking crystallization form, with its centre on the vertical axis (in the S-zone, diagram 2). The individual crystals like fine needles radiate outwards from a point; those radiating toward the periphery have longer rays than those situated toward the centre of the picture. The form stands out conspicuously from the rest of the crystallization type in which a few other unimportant deformations are to be found.

Fig. 39. Photographic enlargement of the typical section of the crystallization.

Diagnosis of Dr. K.:—Prostatitis, inflammation of the prostatic gland.

Fig. 40. Besides numerous other deformations we have a striking form, similar to that in the preceding crystallization plate—a radiation from one point extending outward in all directions. This appears in the fourth quadrant toward the periphery.

Diagnosis of Dr. K.:—Pyelitis (inflammation of the calyx of the kidneys).

We might be in some doubt here as to whether the form indicated refers to the right kidney point or to an inflammatory disease of the right extremity. The solution is given by the presence of the left kidney point, lying somewhat lower down toward the centre and from the fact that the radial lines from the right kidney point do not extend to the periphery, but are definitely broken off near the border zone. In diseases of the extremities the forms nearly always radiate right through to the periphery, thus breaking through the border zone formation. (Compare figures 48 and 54.)

In examining such pictures it is necessary to base the judgment first of all on the characteristic symbol. When various other forms less characteristic are present, i. e. if only one or two-winged crystal centres or broken centres or indiscriminately scattered voids are present, it is the character-

26

istic form which determines the collective significance of the crystallization.

Besides these are some 22 other cases of inflammatory diseases—angina for instance. The centre of this disease is situated in the vicinity of or slightly above the centre of gravity (throat zone). This enables us to establish the fact that the above described form is characteristic of inflammatory processes. In acute stages of tuberculosis this form is frequently to be found intermingled with the Tbc. forms in the blood crystallization picture.

Plate XII

Fig. 41. Case No. 401, female, aged 65. The characteristic centres of this crystallization are to be found in the middle of the third quadrant on the 45 degree radius (gastric zone). It has been possible by special lighting during the photographic process to cause these forms to stand out as darker shapes in the photograph. The centres may be described as lying on a base line drawn through them, on which lie two wings, widening out to the right and to the left from a central point, the crystals not extending very far from this central point. As a rule these forms are soon intercepted by similar forms which cut them transversely.

Fig. 42 shows the enlargement of the typical form, which here is clearly to be seen in detail.

Diagnosis of Dr. K.:—Carcinoma of the stomach in an advanced stage, firmly established anaemia (haemoglobin 19%), during the last few months lassitude, vomiting, loss of weight.

Plate XIII

Fig. 43. Case No. 165, female, aged 61. In the third quadrant, somewhat concentrated near the intersection of the axes, are many of the two-winged centres on a base line as just described. We drew attention in our report on this crystallization that the blood crystallization picture shows distinct cancer signs in the region of the stomach, extending to and including the oesophagus. In the photograph unfortunately this is difficult to see. In the periphery of the fourth quadrant at the right of the vertical axis there is a dark, oblong patch. In the original this is visible as a yellowish-green coloring of the crystals. Such yellowish-green coloring is due to adhesion of water to the copper chloride crystals. It has been found by experience to be a sign of congestion in the digestive tract. The point where this coloring may be observed indicates the region of the liver. We conclude from this that there is a slight congestion of the liver.

27

Diagnosis of Dr. S.:—Tumor of the stomach-orifice and oesophagus, slight congestion of the liver, slight jaundice with bilious attacks. The radiograph shows great contraction of the upper part of the cardiac region.

In 109 other cases where the medical diagnoses refer to cancerous tumors, these characteristic centres were determined indeed with 100% accuracy.

Are we then really justified in regarding these forms as characteristic of cancer and are we as a result of this again confronted with the strange fact that by means of a few drops of the patient's blood these forms appear in the copper chloride crystallization? Here again our collaborator Mr. Rascher in his investigations confirmed our observations.

Fig. 44. This crystallization was obtained by admixing an extract made from a cancerous tumor of the oesophagus with the copper chloride solution. The extract was made from sections obtained after an operation for the removal of the tumor. The two-winged centres are distinctly visible in the crystallization.

Plate XIV

Fig. 45. Extract from the tissue of a cancer of the abdomen removed by operation. Here the two-winged form with one wider, larger wing is especially clear. The extract of the cancerous tissue removed during operation gives the same forms that appear in the blood crystallization plate.

Fig. 46. Case No. 55, female, aged 44. Unfortunately it has not been possible in this case to obtain a sharp, clear photograph, because the whole picture was covered uniformly with encapsulled yellowish-green water. By means of a line in the photograph, we have indicated a two-winged form in the second quadrant near the point of intersection of the two axes. Our diagnosis:—A cancerous form in the region of the throat.

Medical report of Dr. Kn.:—Cancer of the soft palate, operated on several times, also treated with X-rays.

Fig. 47. Case No. 88, female, aged 61. The enlargement of the characteristic area is shown. In the blood crystallization plate this part is to be found to the left of the "centre of gravity". Slightly displaced laterally, in the upper part of the chest, is a distinct cancer indication.

Report of Dr. S.:—Cancer of the left breast, operated, extensive resection.

Further cases of this type are given in chapter III.

Plate XV

Fig. 48. Case 107, male, aged 34. This is one of the most singular blood crystallizations that has been obtained. The third and fourth quadrants are almost normal; on the border between the first and second quadrants there are strong deformations and quantities of encapsuled yellowish-green water. On the left arm of the horizontal axis there is a large void, on the right arm of the axis there is a "corresponding" form. The crystallization picture was extraordinarily difficult to photograph and while the photograph shows the position, it does so rather darkly and indistinctly. A comparison with other cases with similar forms indicates sarcoma.

Fig. 49. Enlargement of the left section of the picture. It is evident that in this plate we are dealing with an injury to a limb (arm, leg), because of the fact that the rays from the characteristic centres extend through the border zone to the outermost edge, thus breaking up the character of the border zone. In such a case we know that it is not an *internal* organ with which we have to deal. Compare this with such forms as are to be found in Figs. 40 or 54, (kidney), where the border zone is distinctly detached. In our report it is noted:—"There appears to be something due to a violent shock or to a serious physical injury."

As in all blood crystallizations we knew nothing about the case before the blood test was examined. Only after making the report the medical diagnosis was consulted. In this manner the crystallization specialist should train himself through pure observation, by comparing already known and examined cases with the present yet unknown i. e. the one under consideration, he should train himself to make a proper diagnosis of the blood crystallization, one that is unbiassed and without prejudice, finally drawing conclusions only from the forms themselves.

This was the procedure in the present case. The subsequent medical report of Dr. S. stated: Patient had, as a child, a sarcoma of the left arm. The arm was amputated, since then naevi on the face, lipoma on the left shoulder and fibroma on the right upper arm have developed.

Mr. Rascher furnished further adequate proofs with preparations made from pathological specimens.

Plate XVI

Fig. 50. An extract from a round-celled sarcoma of the stomach is added to the copper chloride prior to crystallization. We see here the two-winged cancer centre with a large perforated form set directly upon it—the characteristic sarcoma indications.

Fig. 51. Extract from a melanotic sarcoma of the lung. Photographic enlargement of the characteristic zone.

Plate XVII

Fig. 52. Case 8, male, aged 43. We have only had the opportunity of observing very few blood crystallizations of this illness. A more intimate acquaintance with crystallization forms and the comparison with hundreds of possibilities (partial crystallizations) enables anyone who is willing to take the trouble to familiarize himself with this language of form, to make a correct diagnosis even in a case as yet unknown. In the sense of Goethe we may also speak here of the metamorphosis of certain basic forms—archetypes. In this case we had only to observe the crystallization and note that above and below the centre of gravity there are two clear cancer lines. The centre of gravity itself is indicated by a sign more characteristic of tubercular inflammation. Between this and the cancer line is a somewhat unformed "spongy" area. Our conclusion:—a fungating cancerous deterioration of the gland tissue of the cervico-thoratic region with however partially tubercular indications.

Medical report of Dr. S.:—"Hodgkins disease." As is well known this disease—although it has characteristics of its own—resembles tuberculosis as well as lymphosarcomatosis. In the blood crystallization tumorous and tubercular signs are also intermingled.

We will now give a few further illustrations from the numerous individual cases which exhibit characteristic indications.

Fig. 53. Case 194, male, aged 35. The form indicated by an angle in the first quadrant is characteristic of pyorrhoea. The flamelike chaotic formation of numerous centres over the rest of the plate is peculiar to this illness. From several patients differing in age and sex, crystallizations were obtained, almost identical with the present case. Individual characteristics which show themselves in the blood picture of many diseases are not to be found in these crystallizations.

Plate XVIII

Fig. 54. Case 45, male, aged 42. Report of Dr. S.:—Sclerosis of the kidneys. This plate is extraordinarily characteristic and enables us to fix the position of the kidney diseases in the blood crystallization plate. We have marked the areas in question with black points. Even though we have here one-sided radiating crystal-groups extending from these points to the

Plate I.

Photographs 3/5 natural size

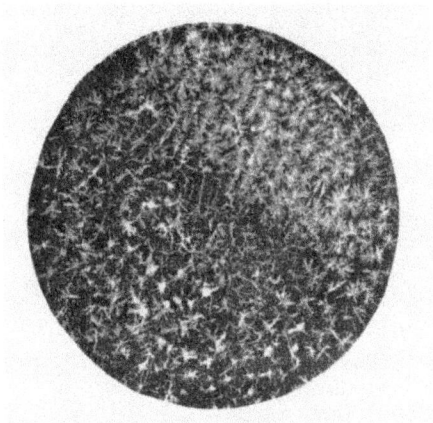

Fig. 1 Cu Cl$_2$ 20%
copperchloride alone
     without any addition
     (transmitted light)

Fig. 2 Cu Cl$_2$ 25%
    + 1 drop of waterlily-flower
     extract

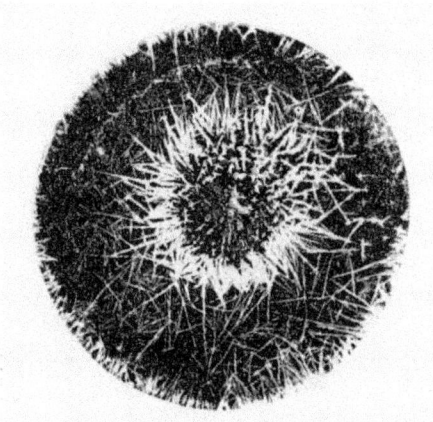

Fig. 3 Cu Cl$_2$ 25%
    + 1 drop of extract from
     agava americana

Fig. 4 Cu Cl$_2$ 5%
    + 1 drop of chamomile-flower
     extract

Plate II.

Photographs 3/5 natural size

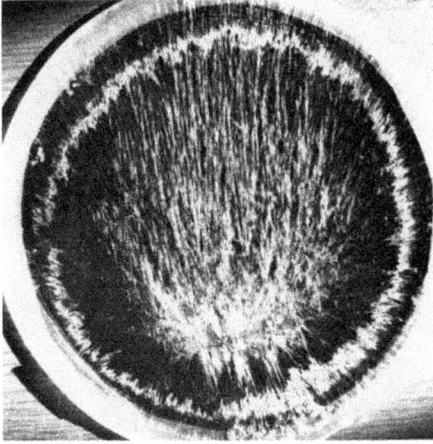

Fig. 5 Cu Cl$_2$ 5%
    + 5 drops of honey

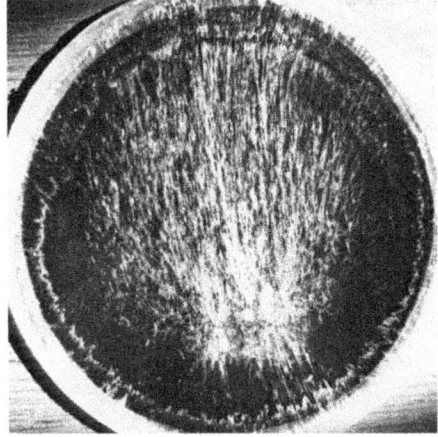

Fig. 6 Cu Cl$_2$ 5%
    + 5 drops of honey

Fig. 7 Cu Cl$_2$ 5%
    + 5 drops of honey

Fig. 8 Cu Cl$_2$ 5%
    + 5 drops of honey

Plate III.

Photographs
1/2 natural
size

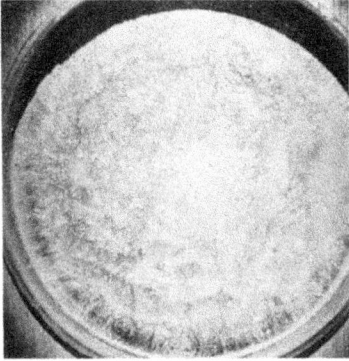

Fig. 9 Copper chloride 20%
without any addition
(reflected light)

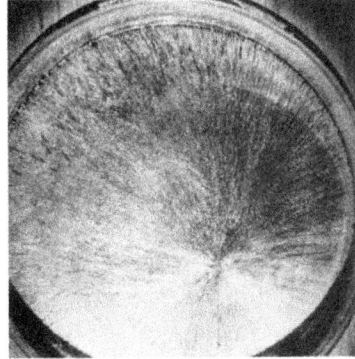

Fig. 12 Copper chloride 20%
+ human blood

Fig. 10 Magnesium
sulphate 20%
without any addition

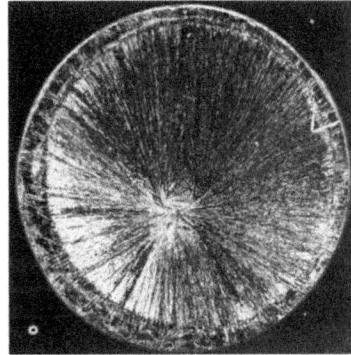

Fig. 13 Magnesium
sulphate 20%
+ the same blood

Fig. 11 Lead acetate 10%
without any addition

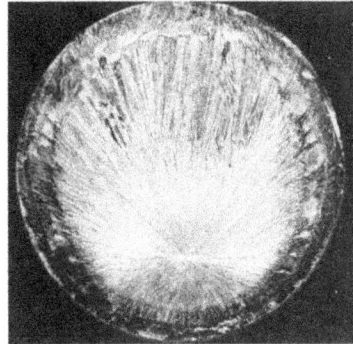

Fig. 14 Lead acetate 10%
+ the same blood

Plate IV.

Photographs
1/2 natural
size

Fig. 15 Cu Cl$_2$ 10%
+ 100 drops
NaHCO$_3$ 10%

Fig. 16 Cu Cl$_2$ 10%
+ 50 drops NaHCO$_3$

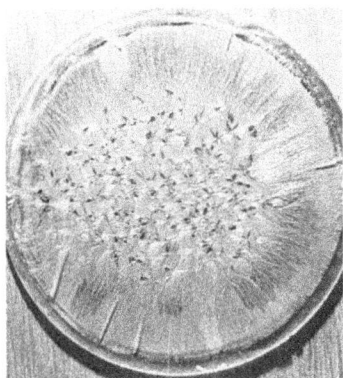

Fig. 17 Cu Cl$_2$ 10%
+ 30 drops NaHCO$_3$

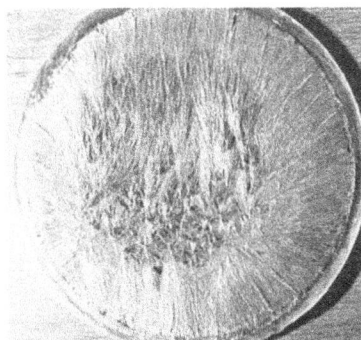

Fig. 18 Cu Cl$_2$ 10%
+ 10 drops NaHCO$_3$

Fig. 19 Cu Cl$_2$ 10%
+ 5 drops NaHCO$_3$

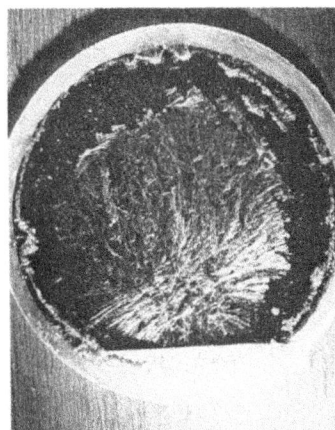

Fig. 20 Cu Cl$_2$ 5%
+ 10 drops of
paraldehyde
10%

Plate V.

Photographs 21 and 22: 3/5 natural size

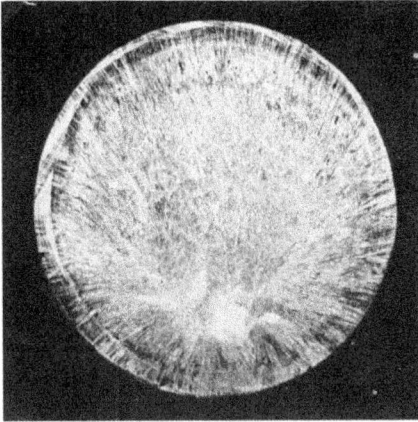

Fig. 21 Cu Cl$_2$ 5%
   + Extract from seeds
   of a pine tree with
   crooked trunk

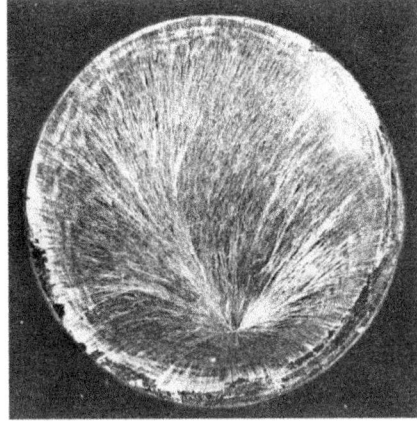

Fig. 22 Cu Cl$_2$ 5%
   + Extract from seeds
   of a normal pine
   tree

Fig. 23 The crooked pine tree

Fig. 24 The normal pine tree

Plate VI.

Photographs 4/5 natural size

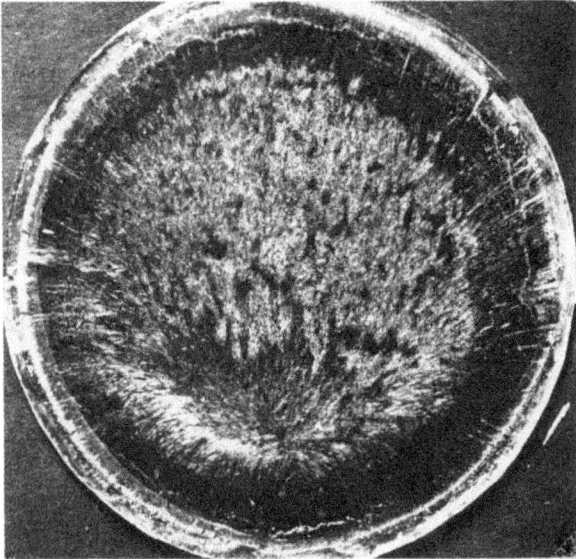

Fig. 25 Cu Cl$_2$ 5%
    + Extract of
    beechnuts from
    a healthy tree

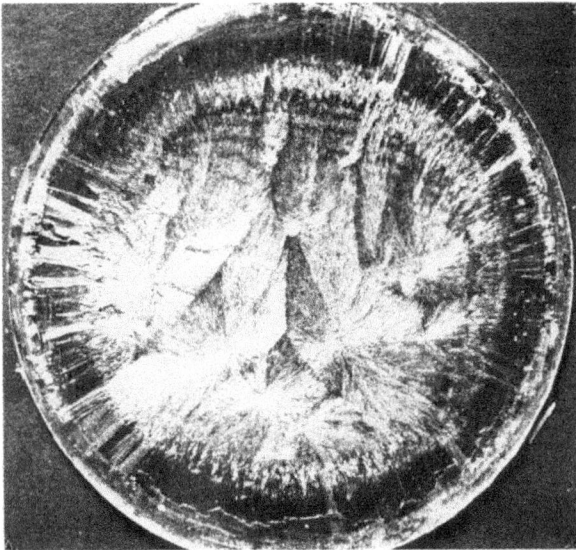

Fig. 26 Cu Cl$_2$ 5%
    + Extract of
    beechnuts from
    an unsound
    tree

Plate VII.

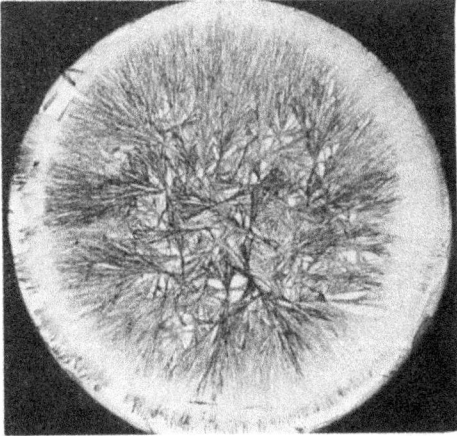

Fig. 27 Cu Cl₂ 20% *
+ Blood from a patient with
tuberculosis of the larynx
3/5 natural size

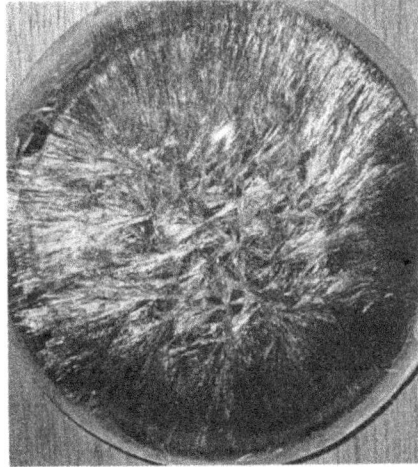

Fig. 28 idem, Different lighting
3/5 natural size

Fig. 29 idem, Enlargement of the typical form
2½ times natural size

Plate VIII.

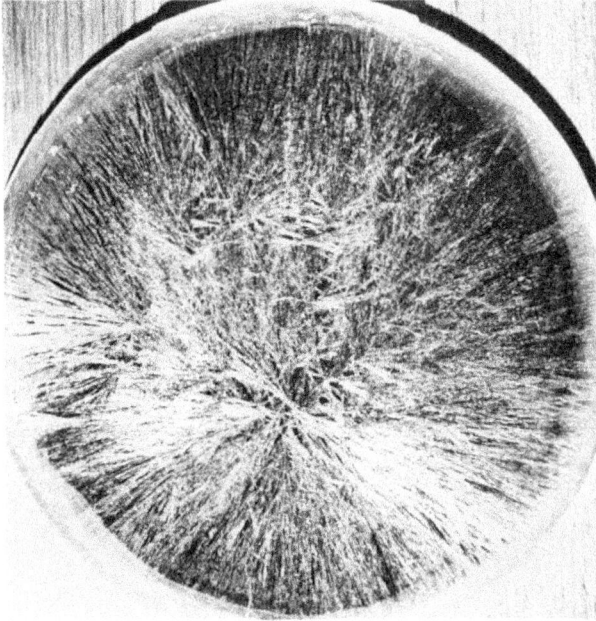

Fig. 30 Cu Cl$_2$ 20%
+ Blood from a
patient with
pulmonary
tuberculosis
9/10 natural size

Fig. 31 idem, Enlarge-
ment of the
typical form
3½ times natural size

Plate IX.

Photographs 3 times natural size

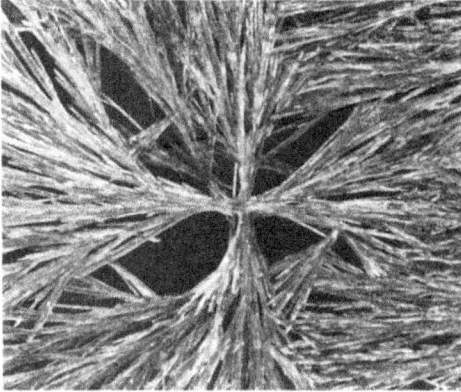

Fig. 32 Cu Cl₂ 5%
    + Extract from a
        pathological specimen of
        a tubercular lung

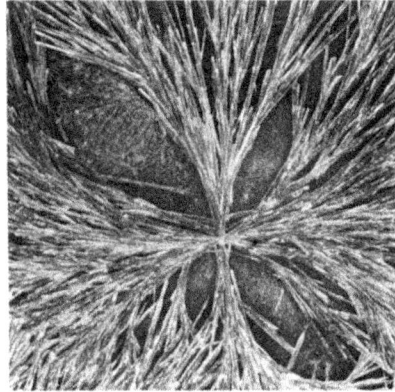

Fig. 33 Cu Cl₂ 5%
    + Extract from a
        pathological
        specimen of a
        tubercular lung

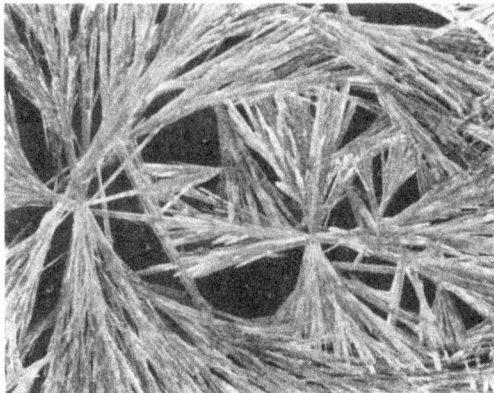

Fig. 34 Cu Cl₂ 5%
    + Extract from a
        pathological specimen of
        a tubercular lung

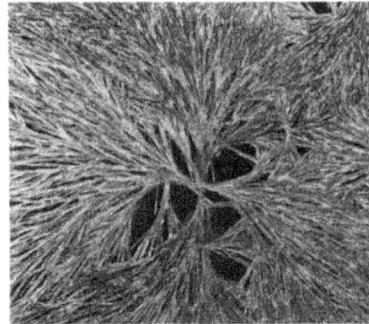

Fig. 35 Cu Cl₂ 5%
    + Extract from a
        pathological
        specimen of a
        tubercular lung

Plate X.
Photographs 8 times natural size

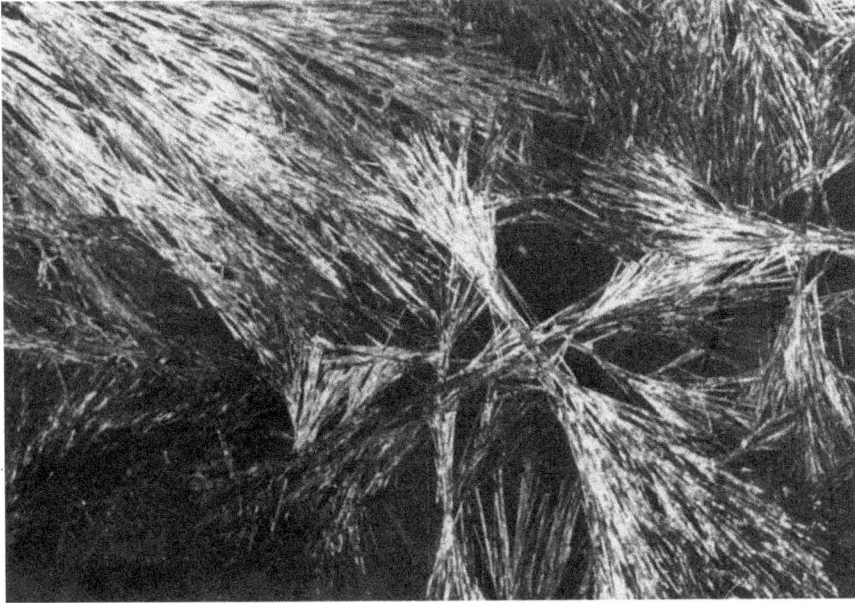

Fig. 36 Cu Cl₂ 20%
+ Extract from a pathological specimen: Tuberculosis of
the right elbow

Fig. 37 Cu Cl₂ 20%
+ Extract from a pathological specimen: Periost-Tbc
capituli ulnae dextrae

Plate XI.

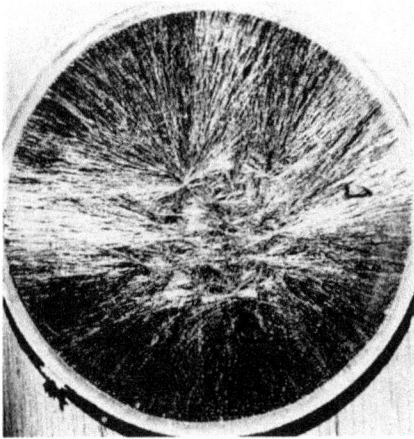

Fig. 38 Cu Cl₂ 20%
        + Blood from a patient with
           Prostatitis
                    3/5 natural size

Fig. 39 idem, Enlargement of the
         typical form
                    3 times natural size

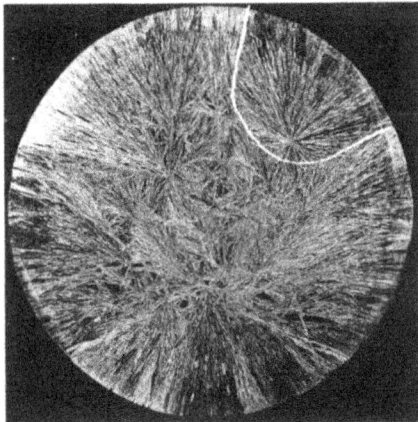

Fig. 40 Cu Cl₂ 20%
        + Blood from a patient with
           pyelitis
                    3/5 natural size

Plate XII.

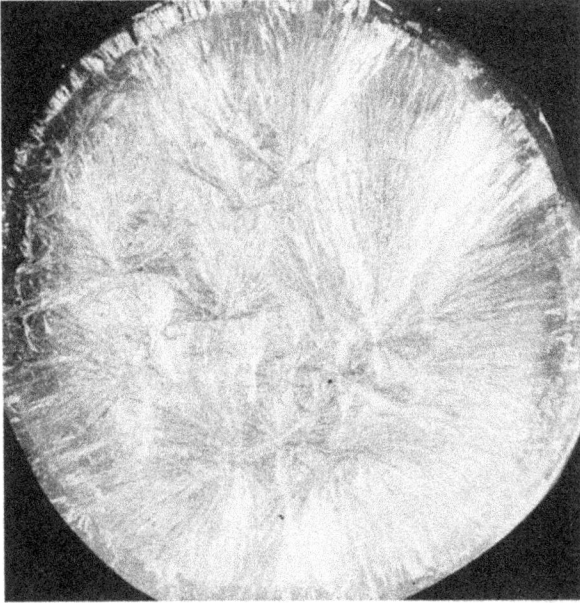

Fig. 41 Cu Cl₂ 20%
   + Blood from a
   patient with
   carcinoma of
   the stomach
   9/10 natural size

Fig. 42 idem, En-
   largement
   of the
   typical
   form
   2½ times
   natural size

Plate XIII.

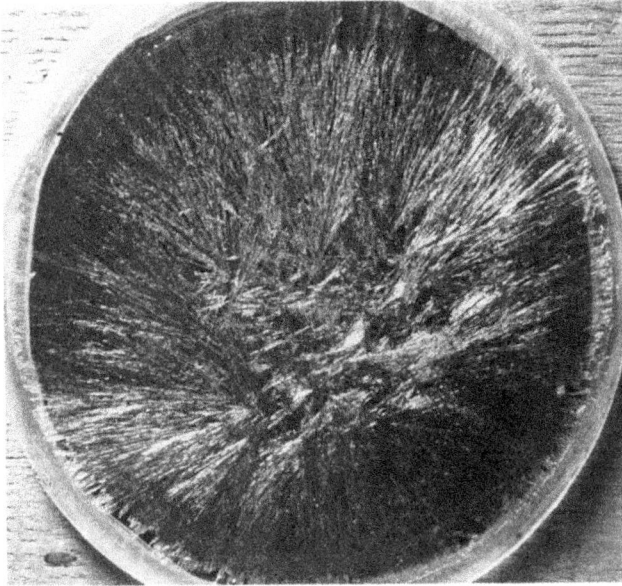

Fig. 43 Cu Cl₂ 20%
  + Blood from a patient with tumor of the
    stomach orifice and oesophagus
                              9/10 natural size

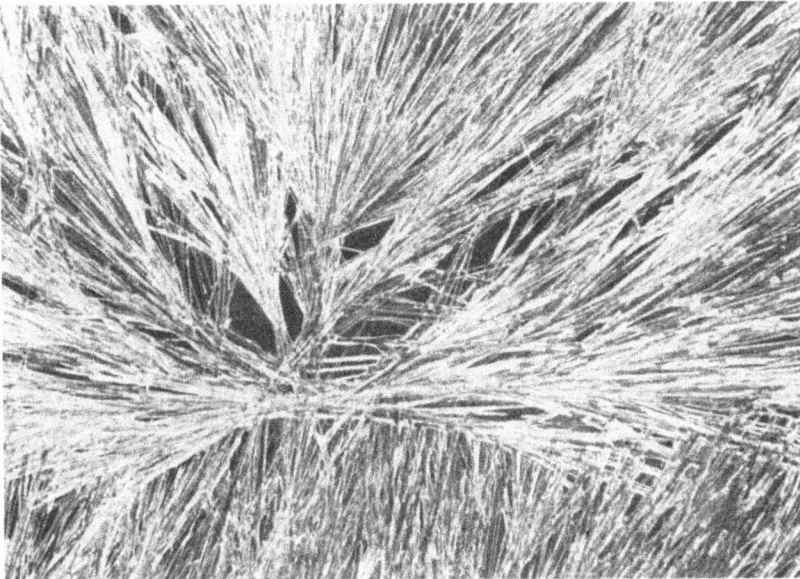

Fig. 44 Cu Cl₂ 10%
  + Extract from a pathological specimen of a tumor of
    the oesophagus
                              9 times natural size

Plate XIV.

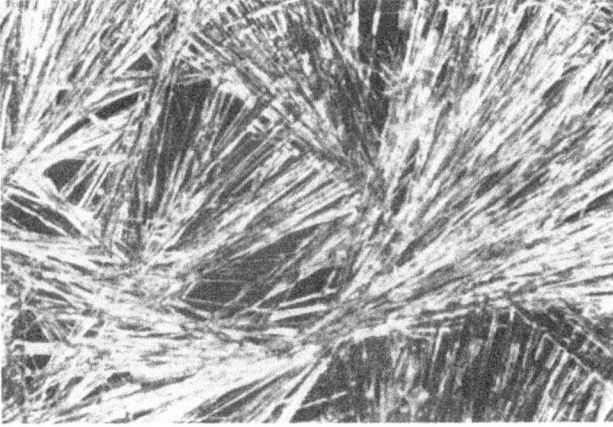

Fig. 45 Cu Cl₂ 10%
+ Extract
from a
pathologi-
cal speci-
men of a
tumor of
the
stomach
9 times
natural size

Fig. 46 Cu Cl₂ 20%
+ Blood of a patient
with carcinoma of
the soft palate
2 times natural size

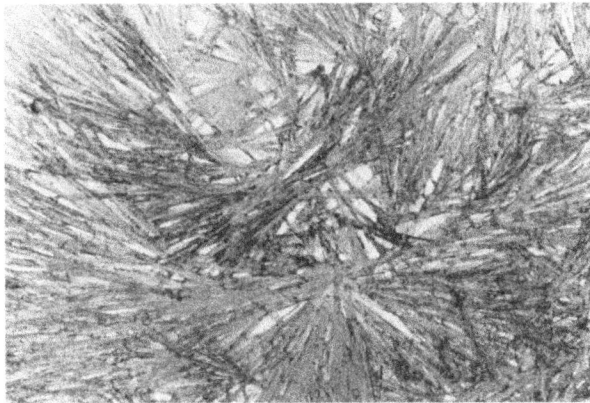

Fig. 47 Cu Cl₂ 20%
+ Blood from a
patient with
carcinoma of
the breast
2½ times natural size

Plate XV.

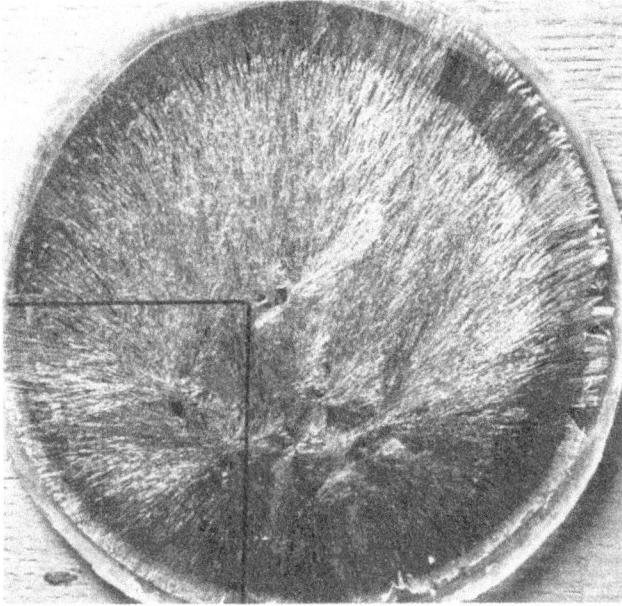

Fig. 48 Cu Cl₂ 20%
+ Blood
from a
patient
with
sarcoma
9/10 natural size

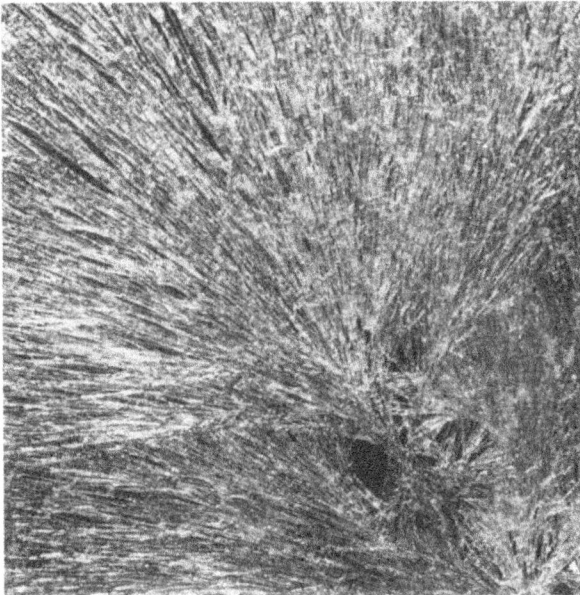

Fig. 49 idem, the left part
of the picture
3 times natural size

Plate XVI.

Fig. 50 Cu Cl₂ 10%
        + Extract from a pathological specimen of a round-
          cellsarcoma of the stomach

                                    10 times natural size

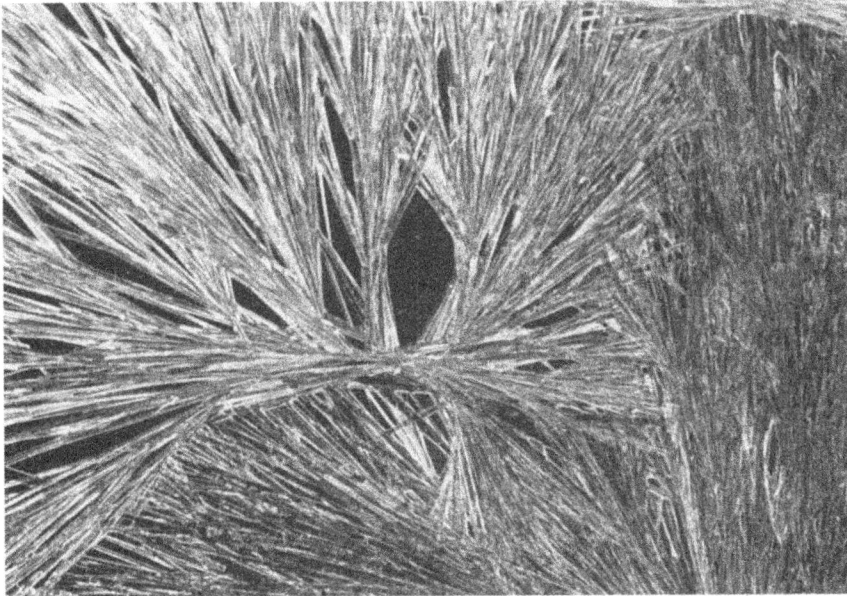

Fig. 51 Cu Cl₂ 10%
        + Extract from a melanosarcoma of the lung

                                    10 times natural size

Plate XVII.

Photographs 9/10 natural size

Fig. 52 Cu Cl$_2$ 20%
+ Blood from a
patient with
Hodgkin's
disease

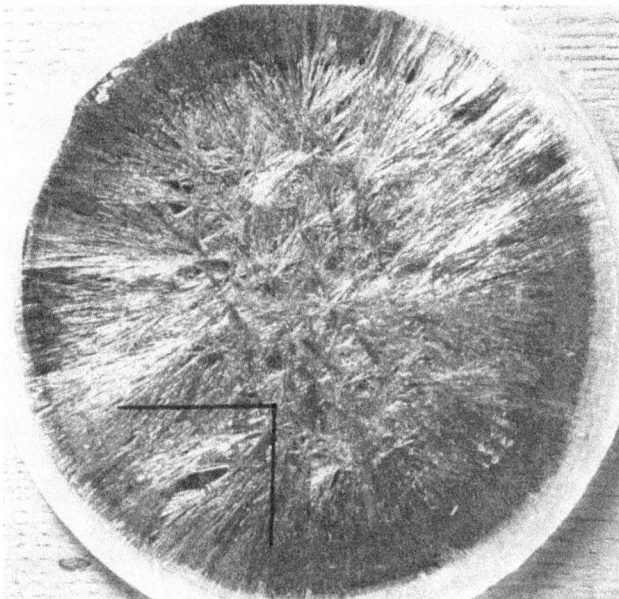

Fig. 53 Cu Cl$_2$ 20%
+ Blood
from a
patient
with
Pyorrhoea

Plate XVIII.

Fig. 54 Cu Cl₂ 20%
+ Blood from a patient
with sclerosis of the
kidney
9/10 natural size

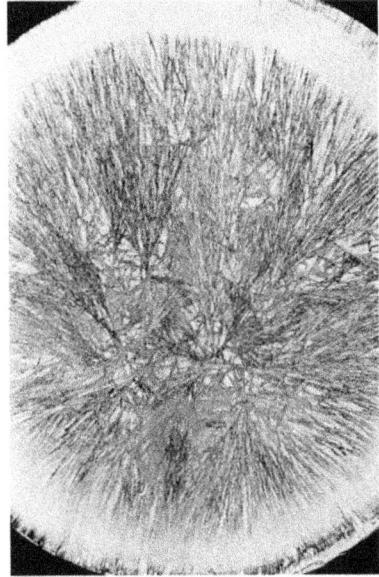

Fig. 56 Cu Cl₂ 20%
+ Blood from a patient
with calcified
tuberculosis
9/10 natural size

Fig. 55 Cu Cl₂ 5%
+ Extract from a pathological specimen of Carotid
Sclerosis

Plate XIX.

Photographs 4/5 natural size

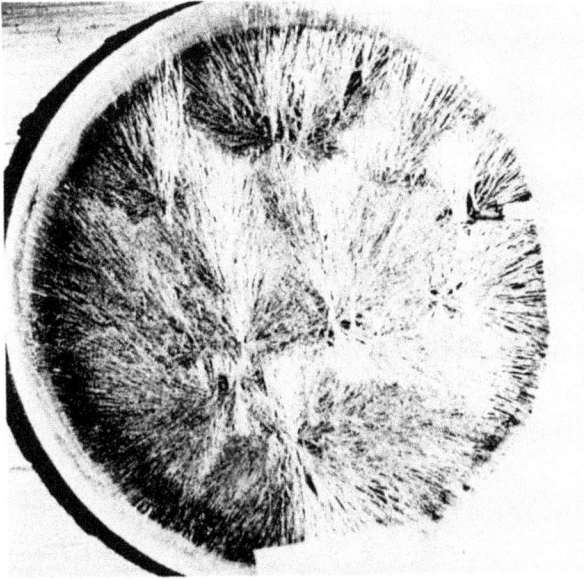

Fig. 57 Cu Cl$_2$ 20%
+ Blood from a
patient with
carcinoma of
the rectum

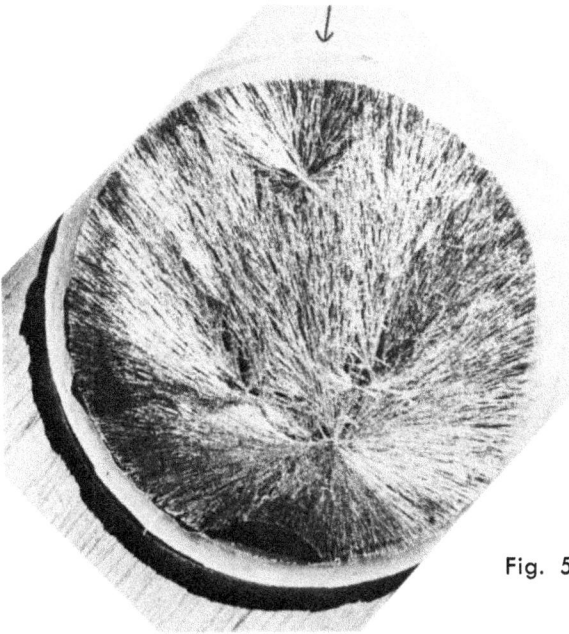

Fig. 58 idem, Blood
+ Viscum abietis Z 28

Plate XX.

Photographs 4/5 natural size

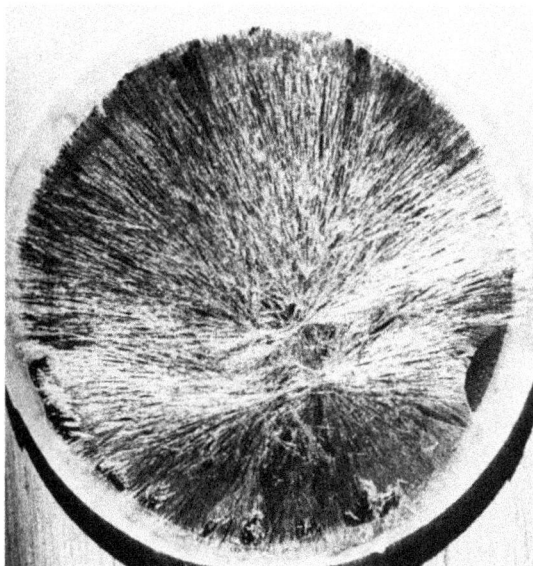

Fig. 59 Cu Cl$_2$ 20%
+ Blood from a
patient with
carcinoma of
the larynx

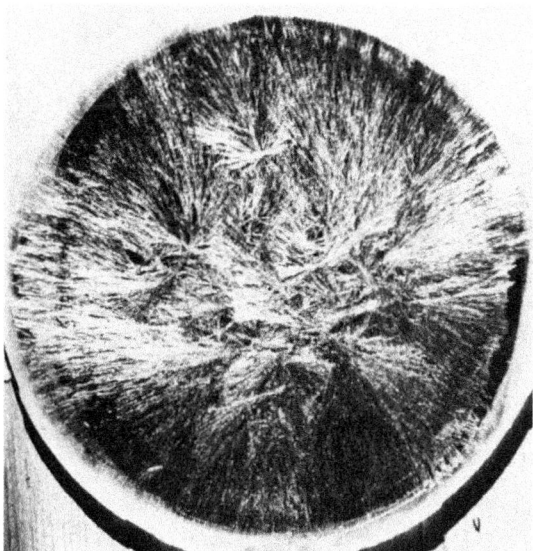

Fig. 60 idem, Blood
+ tomato juice

Plate XXI.

Photographs 4/5 natural size

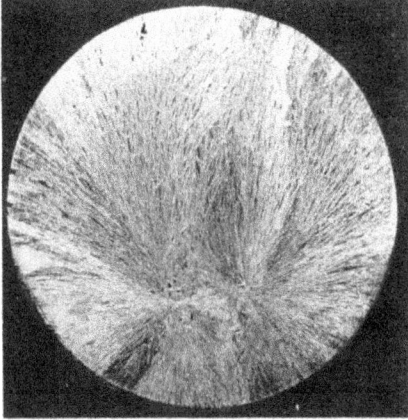

Fig. 61 Blood from fig. 41
        + Urtica dioica D 3

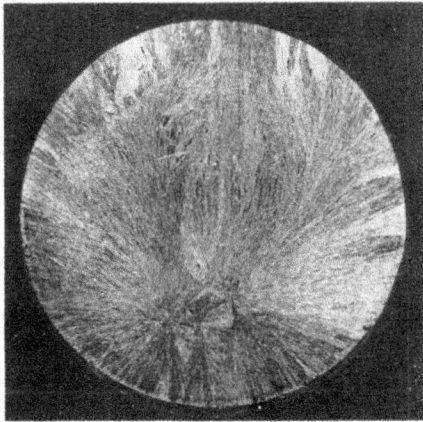

Fig. 62 idem, Blood
        + Iscador Z 7

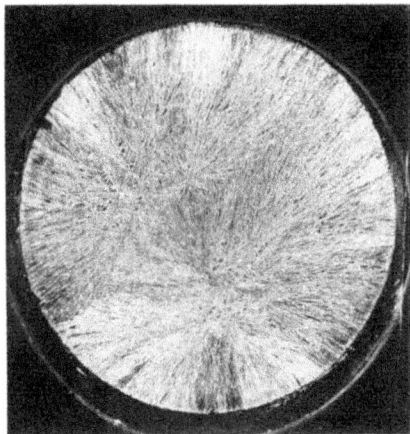

Fig. 63 idem, Blood
        + Helleborus niger Z 14

Plate XXII.

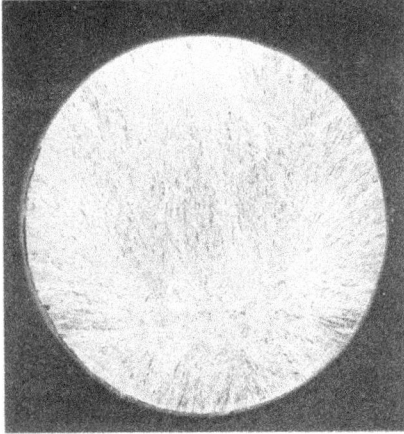

Fig. 64 idem, Blood
+ Antimony D 6
4/5 natural size

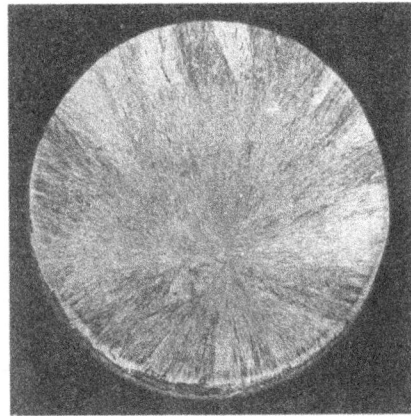

Fig. 65 idem, Blood
+ Iscador Z 7
+ Urtica dioica D 3
4/5 natural size

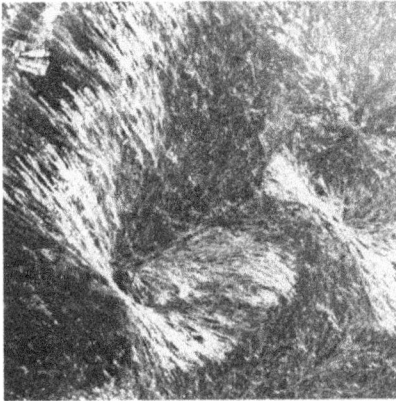

Fig. 66 Cu Cl₂ 5%
+ Extract of Viscum
mali
Enlargement of the
typical form
2½ times natural size

Plate XXIII.

Photographs 4/5 natural size

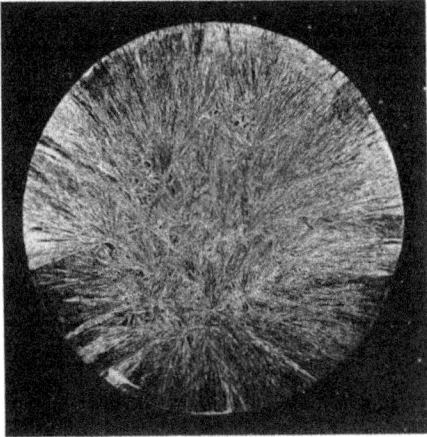

Fig. 67 Cu Cl₂ 20%
   + Blood from a patient
     with asthma

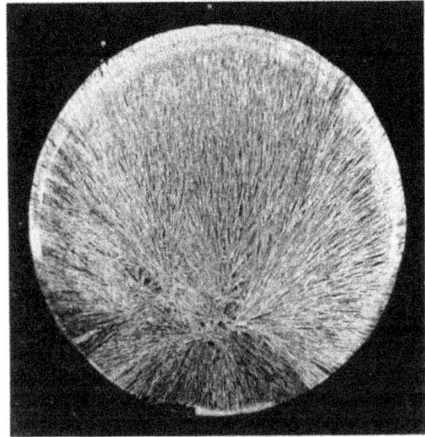

Fig. 68 idem, Blood
   + Gencydo 1%

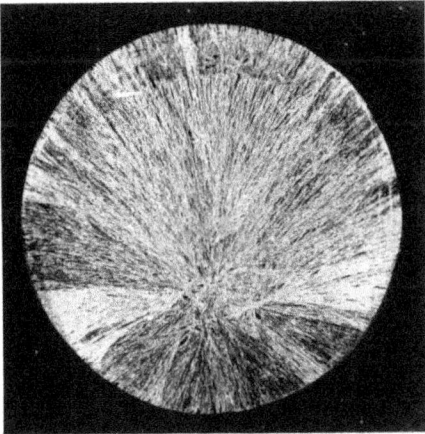

Fig. 69 idem, Blood
   + Prunus spinosa

Plate XXIV.

Fig. 70 Cu Cl₂ 5%
+ 15 drops Retina D 2
5½ times natural size

Fig. 71 Cu Cl₂ 20%
+ Blood from a patient
with detachment of
the retina
7½ times natural size

Fig. 72 Cu Cl₂ 10%
+ 15 drops of extract
of Pelvis renalis
D 3
Double natural size

Plate XXV.

Fig. 73 Cu Cl$_2$ 5%
+ 1 drop Belladonna Urt.
6 times natural size

Fig. 74 Cu Cl$_2$ 20%
+ Blood from a
patient,
showing
belladonna-
forms
6 times natural size

Fig. 75 Cu Cl$_2$ 5%
+ 7 drops Medulla spinalis
D 2
2 times natural size

Plate XXVI.

Fig. 76 Cu Cl₂ 5%
+ 5 drops Medulla
oblong. D 2
natural size

Fig. 77 Cu Cl₂ 5%
+ 5 drops
Medulla
oblong. D 2
5 times natural size

Fig. 78 Cu Cl₂ 5%
+ 7 drops Corpus
quadrigeminum
D 2
1½ times natural size

periphery in the third and fourth quadrants, it is not the true indication of arterial sclerosis, but rather of a functional obstruction.

Fig. 55. A plate obtained by Mr. Rascher. This shows the characteristic zone of the blood crystallizations of an organic preparation, made from a post mortem dissected section taken from an organ of a female, aged 65, suffering from sclerosis of the carotid arteries.* The large void (perforation) which occurs in the curved crystallization groups assumes a sort of gothic arch-like form which is characteristic of the true arterial sclerosis. Besides the diseases mentioned here, 60 cases of nervous and mental diseases, 45 cases of diseases of the digestive tract and 150 cases of other diseases were investigated and diagnosed with an open mind.

Exact knowledge of the typical forms and of the modifications of the whole plate due to the appearance of a particular centre is of the greatest importance for the diagnosis of individual blood crystallizations. It has already been shown by several examples that the particular grouping of the different kinds of centres permits the location and the determination of the position of the diseased organ. We wish to recall to your mind that the characteristic zone for the throat is at the centre of gravity of the crystallization plate, for the stomach it is in the inner angle of the third quadrant; for the lungs it lies right and left of the A-axis somewhat below, and for the liver it is somewhat away from the centre in the fourth quadrant, yet more in the neighborhood of the M-Axis. If greatly deformed centres appear in the liver zone, then all other forms are frequently displaced to the left, in the direction of the upper angle of the second quadrant. The stomach zone lies then more toward the periphery. It is difficult here for these cases to give exact points, because every case is individual. Nevertheless the practised observer can recognize the displacement possibilities from the relationship and the equilibrium of the forms. An extensive perforation of the lung zone for example penetrates deeply right into the lower structure. Metastasis in the stomach zone accumulate repeating forms right out into the periphery of the third quadrant.

The imprint of the diseases of the outer extremities is to be found in the forms which have been pushed from the periphery of the border zone right into the innermost region of the quadrants.

A case of orderly displacement takes place somewhat toward the left in the third quadrant in the region characteristic of the rectum. The kidney points do not lie so low in the neighbourhood of the centre of the plate in the case of the male as in the case of the female organism. They are often

---

* The detailed reports of the Pathological Institute of the University of Basle concerning this and the other specimens obtained by Mr. Rascher may be examined in our files.

only to be recognized by indications right and left of the A-axis, mostly as somewhat displaced symmetrical formations. The ovary points are to be found a little "below".

We discover in this way a certain "overlaying" of disease indications— for example—of the bladder and the uterus, of the mammae and the lungs. Which of the named organs is sympathetically affected (synalgia) can only be judged from experience and the comparison of "possible" forms. For this reason it is especially difficult to judge the head zone, because in this region points of varying character are compressed together in the small space of quadrants 1 and 2. Here often only the individual type of the centres can elucidate the problem.

Primarily there are two chief types. One shows a crystallization form with spoke-like radiating crystals proceeding from one or more centres. The individual lines of the crystal are clearly detached from one another, the rays proceeding more or less in all directions or in bundles from the centre under consideration. The clearest case of this kind is the inflammation centre, to be seen in Fig. 38 and 40.

Such crystallizations show also a bright glitter overspreading the crystals; these are transparent, the interspaces clear. This is the inflammation type, in other words, indications of increased metabolism, increased blood activity and increased disintegration. If the ray bundles are crudely formed with thicker and fewer rays, that is to say "of watery appearance" of a somewhat yellow-green color, then we can recognize from the lack of characteristic centres, such as the Tbc. forms or the inflammation forms, the presence of a general superacidity of the organism in connection with disturbance of the metabolism. In cases of rheumatism such "radiations" appear proceeding from the border zone toward the centre, in most cases surrounded by encapsulated yellow-green liquid.

If these ray-like structures appear finely membered, almost silky with a special shimmer and if they are thickly pressed together, we can conclude that they arise from pure nervousness or even from psychic symptoms. In epileptic children patients such a variation has been observed in a few cases.

In three cases of schizophrenia (split personality) I noted that the radiations proceed from two centres dividing the blood crystallization into two completely independent parts. As a matter of fact in these cases we were not able to obtain sufficient material for observation in order to discover other characteristics; nevertheless, I should like to remark here, with the admixture of pathological preparations very characteristic variations were established. And in two cases of schizophrenia a high potency of a heart-preparation, D 30 (lower potencies had no effect), was admixed with the diluted blood of the patient in question; a "normal" crystallization was the result.

In a few cases of a condition of maniacal depression the admixture of liver preparations had the effect of a positive reaction and in the case of mere depression the admixture of a lung extract D 30 produced a distinct result.° Another fundamental crystallization type is one in which the crystals lie closely alongside one another without any radiating arrangement. Densely piled up crystal layers with little or no sheen result in a form of a dull, opaque character. The centres which appear have short wings often broken off and they seem to be continually disturbed and transected in the process of formation. Large perforated forms indicate that intense organic injuries to the body are already in process. This is the type of obstructed processes, of tumification leading to carcinoma and sarcoma forms. If such densifications seem only to be local, appearing as extraneous bodies in the remainder of the crystallization plate, then we can conclude that there are hardening processes taking place, especially frequent in the zone of the liver and the kidneys. Combined with open spaces, the enframing of which shows curved crystal shapes, it indicates sclerotic processes. (Compare Fig. 55.)

As already mentioned, the formation and breadth of the border zone informs us of the extent of the vitality. The rays extending from the centre of gravity can continue right out to the periphery and show a uniform structure. The crystallization picture becomes finer and finer at the periphery. It can be compared to the fine distribution of a capillary system. We have the same type when we think of the circulation of the blood and note how the arteries branch out more and more. This was to be seen most impressively in some remarkably prepared anatomical embryo specimens where the skulls were made transparent, so that it was possible to see the fine branching out of the veins. The laws of form-structure which confront us here are quite general for the whole of nature wherever the form-giving activity of liquids, water, etc., becomes operative within solid or semi-solid materials. Aeroplane photographs of landscapes with river courses—which branch out continually finer and finer into riverlets and brooks lying in the mountains—show in a striking way a similar arrangement. Even in aeroplane photographs of eroded valleys in desert regions, we find similar indications.

In general every tree shows the same character when we follow the course of its vascular tissue, from the stem into the branches, into the

---

° We employed for this preparations of the Weleda Co., Arlesheim, Switzerland.

smallest twigs, right out into the finest divisions of the leaves. Here we have a concentric gathering together toward the middle in the stem and the fine branching out and division toward the periphery into the "border zone of the leaves". We are now confronted with a completely analogous law of form. The healthy tree has a perfect formation of this border zone in the most delicate arrangement. Whoever has studied a tree in winter, will have already observed this. In the symmetrical or distorted division of the finest twigs round the crown of an unconfined tree, we can actually read its condition of health. Dried out twigs with numerous open spaces indicate disturbance in growth which leads back right into the activity of the roots. I should like to say, if a personal remark is permissible, that the observation of trees in winter was a part of my "education" for the purpose of training my eyes to penetrate into the particularities of the crystallization formations.

The extremes may be presented in two life drawn types:

Drawing 3a   Drawing 3b

Drawing No. 3a shows a beech tree standing at its full growth. The border zone shows a completely harmonious form, chiselled out into the finest details. It is a picture radiating with life.

Drawing No. 3b shows an old oak, torn by the storm, in a state of decay. Only the branches of the middle region and the stem preserve their existence. The border zone is undeveloped, ragged and gnarled. A picture of decay, of physiological aging.

In a figurative sense we find these two extreme types in the border zone of every blood crystallization. The latter results in a zone in which crystal incrustations are situated between the frequent voids, this is to be seen for example in Fig. 41 or 59.

The appearance of the typical cancer forms has already been described. If a majority of the repetitions of the same form appear together, with "voids" or a wide border zone, this points to a much worse condition of the case in question. If the cancer forms appear in a blood crystallization mixed equally with the Maltese Cross forms, but not markedly indicated, then this is really a contradiction, for in most cases the human being does not suffer at the same time from cancer and tuberculosis. In such cases it has been found that there is no organic condition present, but that it is a question of typically nervous or neurasthenic states. In the case of old people we find from time to time—disregarding the appearance of the "acute" indication in the lung zone—"hardened, atrophied", that is to say, tubercular, one-sided developed forms with *very* short wings. This points to encapsulated, calcified, tubercular conditions from an earlier life period.

Plate XVIII (Continued)

Fig. 56. Case 281 shows in the designated places such indications. We have to deal here with encapsulating processes which had occurred 20 years previously and which even to-day appear in the blood crystallization plate. In this crystallization the kidney and ovary points are also clearly indicated.

This sort of determination, which very often can be made, belongs to the most surprizing results of the methods we are employing. We were compelled to draw this conclusion from our observation of the blood of the patient at the present time. These and other observations show clearly that there are present *real formative forces*, of which the blood is the carrier. These are the forces which are effective in the structure and the constitution of the human being, at the same time causing the differentiation in organs and processes. These forces live themselves out in the bodily nature of the human being in such a way, that they direct completely the catabolic and anabolic processes.* In the crystallization plate of the copper chloride these forces transmitted by a couple of blood drops can naturally only influence the arrangement of the crystal groups, thus revealing the "dynamic" of the expending organization. This is what Rudolf Steiner meant when he spoke of the etheric formative forces.

Goethe did not without intention write the words: "Blood is a very special fluid". It is that bodily fluid, out of which after all the whole human bodily nature is formed. It contains and transports not only *all* the substances which we need for our upbuilding, but it is also the bearer of the

---

* I should like to draw attention here to my book "Crystals" and to my article in the Gäa Sophia, 6th volume, Goethe-Year-Book, entitled "Biological Thinking" as well as to the book of Dr. G. Wachsmuth: "The Etheric Formative Forces in Cosmos, Earth and Man."

inner life-dynamic of our Ego-hood which constructively and destructively takes possession of our organization. These forces are not dependent upon the *individual* chemical molecule. The blood in its entirety is that fluid by means of which—like guiding hands—these forces can seize upon the material substance of the bodily nature.

It is only the human being who recognizes the possibility of an inner dynamic in the upbuilding of an organism, who will be able to develop himself through this knowledge of the dynamic into a skilled, factual interpreter of the crystallization plates. Those who classify the indications mechanically and carry them back only to partly unknown chemicophysical influences, will not develop the activity of perception necessary to review the life processes in a practical way. The perception of the true nature of the human structure and the functions of the organism presupposes also a correct judgment of those forms which can be observed in the crystallization image.

Schelling once said, "to know nature is to create nature". The researcher who experiences—in union with the creative formative forces—those tensions and relaxations which are necessary in order to determine a crystal group in this or that place, imitating with creative intuition the processes of nature, comparing with it his perception of the formation, of the physiological upbuilding and the nature of the human being, for him these crystal pictures begin to speak. He is the one who can "read" them.

What modern science and the research of physics and chemistry have to say about the crystallization and the condition of the blood in this regard is not worthless. It is only one side of the processes of nature, the other side of which must be supplemented by dynamically directing forces in order to form an image of the whole.

Creative intuition* is after all a quality which every doctor and natural science researcher should respect. Confronted by the sick, this is the quality which should lift him from his ordinary knowledge to real knowledge, to wisdom, and from a "craftsman", transform him into an "artist" ** in his profession.

What Goethe named "perceptive power of thought" is the first step toward this quality.

---

* What I understand by "intuition" in this connection is the complete and conscious union with the subject of cognition.

** This word is not used in a critical and prejudicial sense, but only for the purpose of differentiating it from mere technical achievement.

III. Further specific Crystallization Research
— Remedies and the course of Diseases —

It has been shown in the preceeding chapter that every admixture of an organic substance is capable of exerting an influence on the process of a copper chloride crystallization. It is quite understandable that when two different substances are mixed together and then added to a $CuCl_2$ solution, reactions can overlap and exert an influence on each other. The resulting forms may vary widely, from a very slight influence, to one mutually compensating, or even to a continually increasing disturbance. If for example we mix the diluted human blood with plant extracts or extracts made from bodily organs and with this admixture make a $CuCl_2$ crystallization, we shall have as a result very different kinds of reactions.

If we compare these reactions with the blood crystallization of a patient which contains no admixture, we may find the result to be a harmonization*, a so-called "normal" picture may appear (See illustration Fig. 12 plate III). On the other hand the crystallization may remain unchanged or it is even possible that more centres and deformations may appear than were present in the blood crystallization without admixture. From this we may conclude that the formative tendency of the admixture has an influence on the formative tendency of the blood. In order to study this problem more closely, blood tests of a patient were made. A series of crystallizations of his blood were made with admixtures of various plant extracts, remedies or extracts made from organs. The results showed that certain admixtures yielded specific reactions according to the nature of the disease.

We offer here several test examples, after which we shall discuss the problem further.

---

* By this word we mean that the whole crystallization picture tends toward an ideal type which we call "the normal picture". See p. 21.

Plate XIX

Fig. 57. We note quite at the further end (S-zone) of the A-axis one of those two-winged forms, characteristic of carcinoma. The remaining centres of this crystallization plate are less characteristic; they have partly radiating and partly cruciformed structures. In the crystallization pictures of carcinoma patients we notice frequently—as in the present case—a dense, almost opaque structure in the crystal formations. We have thus two primary indications for carcinoma; we may consider the formation below the M-axis as a secondary characteristic which may seem perhaps to contradict the other two. The chief centre of attention is the already mentioned carcinoma formation. The location of this formation is found empirically to be exactly characteristic of diseases of the lower part of the trunk:— rectum, bladder, male and female sex organs. The present case is that of a fifty-six year old man. In diseases of the bladder, delicate crystal radiations are to be seen which extend a third of the way along the A-axis,—these are not present in this crystallization. In cases of diseases of the sexual organs, an area to the right and the left of the A-axis, proceeding from the margin well into the middle of the picture, is clearly demarcated like a segment. These characteristics are also not present, so that we are able to decide with certainty that it is carcinoma of the rectum.

Diagnosis of Dr. K.:—Advanced rectum carcinoma, operated upon several times and radiated. Patient has robust constitution.

Fig. 58. The characteristic reaction is obtained here when a few drops of a preparation of viscum abietis (Z. 28) is mixed with diluted blood and allowed to crystallize. The influence of this viscum preparation has transformed the entire crystallization formation. In the same place where we previously found a carcinoma centre clearly indicated, we now find a clearly defined centre of inflammation i. e. forms which radiate in all directions from a single point.

It is not the purpose of this book to discuss viscum or any other therapy for this or that disease. We only wish to show that with certain remedies very essential changes can be produced. We find these changes of form very significant in recognizing the nature of the disease appearing in the blood crystallization. As a result of numerous experiments it has been observed that in many cases, though not all, the addition of a viscum preparation produces this characteristic change. In the present case after giving the patient an injection of the viscum preparation a rise of temperature was produced. In order to avoid misunderstanding, it should be said that the crystallization containing the viscum preparation was made at the same time as the ordinary blood crystallization itself. Thus there is no question of a subsequent reaction of the patient but of using the same blood for both plates, Fig. 57 and 58.

Plate XX

Fig. 59. Two clear carcinoma forms exactly at the centre of gravity of the crystallization plate. From the first case of Tbc. (Fig. 27) we know that the centre of gravity corresponds to the location of the larynx. The obvious conclusion is:—laryngeal carcinoma. In this plate the broad peripheral zone is of further interest, it indicates great weakness. With such a broad zone—as in this case—we would say that the patient is in a dangerous condition.

Report of Dr. K.:—Carcinoma of the larynx of a 65 year old man, ill for a number of years, several operations and rayings without result. Patient speaks in a broken whisper; the laryngoscope shows growth at the entrance of the larynx; the vocal chords are destroyed; the patient is in a state of suffocation and breathes with difficulty.

In the preceding case (Fig. 58) the addition of the viscum abietis preparation has produced a distinct harmonization of the crystallization. Some centres have completely disappeared as a result of the admixture; the others are drawn together more toward the centre of gravity. This type approaches more nearly the *normal* blood crystallization. The most important carcinoma centre is replaced by an "inflammation form". This result we call a "positive reaction". It is of importance in this case, because not only was there a distinct sign in the S-zone but there was also a contradictory indication present in the raylike form near the M-axis in the third quadrant. It definitely verifies the report.

It is also possible to obtain a "negative reaction". In cases of cancer fresh tomato juice produces this. A few drops of tomato juice added to the diluted blood has in known cases of carcinoma the effect of increasing the number of carcinoma forms in the blood crystallization. This is clearly to be seen in Fig. 60.*

Plate XXI

The following series should be compared with Fig. 41 and 42 which show the ordinary blood crystallization of a carcinoma of the stomach. The experiment was made by mixing the blood with various preparations.**

---

* We do not intend to infer from this that the eating of tomatoes produces cancer.
** For these and the following experiments, the Weleda Co. of Arlesheim, Switzerland, kindly placed some of its preparations at our disposal.

Fig. 61. Blood of the patient (belonging to Fig. 41) mixed with urtica dioica D 3. An extensive unification and harmonizing of the crystallization has taken place. Positive reaction. We call such crystallizations "curative reaction".

Fig. 62. A mixture of the same blood with iscador Z 7. This crystallization also shows—in comparison with the ordinary blood test (Fig. 41)—a decided improvement.

Fig. 63. A mixture of the same blood with helleborus niger Z 14. In comparison with the two preceeding crystallizations this picture is of little value. Increased number of centres, incrustation, etc., negative influence.

Plate XXII

Fig. 64. This crystallization is likewise very unsatisfactory. Numerous incrustations and centres show the negative formative influence of a preparation of antimony.

Fig. 65. A mixture of iscador Z 7 and urtica dioica D 3 produces a unification of the centres which in the blood crystallization (Fig. 41) are distributed over the whole surface of the plate. This is the most favorable reaction. These counterreactions can in practice be used both as aid for diagnosis and for therapeutical studies. Without entering further into the nature of the viscum therapy[*] I should like nevertheless to offer an interesting illustration:—

Fig. 66. Extract of mistletoe, taken from an apple tree, picked in December. The photograph shows the characteristic part enlarged. A few drops of the freshly extracted juice were mixed with a 5% copper chloride solution. In this crystallization it happens that the special characteristic centres which are distributed over the whole surface of the plate are similar in character to the form which has already been determined as being typical of carcinoma centres:—two-winged forms, a short and a broad wing on a base line.

Formative forces similar to those active in carcinoma are thus also found to be active in the mistletoe extract. This connection suggests an explanation of the neutralization and obliteration of the carcinoma forms by the addition of viscum extract to the blood (of Fig. 57/58 and 41/62). This results in the so-called "cure picture".

---

[*] Compare here W. Kaelin: "Die Viscumbehandlung des Karzinoms." Hippokrates, Zeitschrift für praktische Heilkunde, Stuttgart 1933, No. 10.—W. Kaelin: "Viscumprophylaxe des Karzinoms. Frühdiagnose mittels der kapillar-dynamischen Reakion." Hippokrates 1934, No. 2.—Bernhard Aschner: "Die Krise der Medizin. Lehrbuch der Konstitutionstherapie." 1st edition 1928.—A. Neumann: "Die Krebsbehandlung in der täglichen Praxis." 1935.

Plate XXIII

Fig. 67. The blood test of a male asthmatic patient, taken during an attack. The chaotic confusion of the centres is characteristic of a spasmodic condition. Here it is to be seen that a 1% gencydo preparation has an extraordinarily strong form-producing effect upon the blood crystallization. See Fig. 68. A remarkable harmonizing and unifying of the picture is apparent when contrasted with the confused forms of the previous illustration (Fig. 67).

Fig. 69. An admixture of prunus spinosa also produces a clearly harmonized crystallization picture. Both cases show a positive reaction.

Such examples can be multiplied indefinitely, but lack of space has obliged us to limit ourselves to a few illustrations. For the research physician the practicability of this method offers an extensive field of study of the formative forces and of the influence of various remedies on the blood crystallization. In connection therewith it permits us to compare this picture with the actual course of the disease and recovery.

It must be emphasized here that this method is not properly fitted, nor has it been worked out for that purpose, as is often popularly misunderstood, for "experimenting recklessly" in order to discover accidentally what remedy should be used in a particular case. Likewise, the diagnosis and other medical knowledge of the physician cannot and should not be dispensed with, as a result of this method. These remain naturally at the foundation of his work. However just as in medicine various reactions are employed, for instance the Wassermann test, or the reaction to sugar with the Fehling solution, or the blood count etc., etc., so can this "reaction of formative forces" also be used in order to support and strengthen ordinary diagnosis. It is not however at all a universal method but, as in the case of all serious investigation in this field it can become in the hands of an experienced physician a helpful assistant. This ought to be especially accentuated here because I frequently am confronted with the illusionary belief of others who say, "Now we can dispense with the experience of the physician"! But this is a false point of view. Keen insight and medical experience will always be the foundation upon which he stands and such they will always remain. As a matter of fact anyone who has occupied himself in studying the fine form variations of this crystallization method and who has studied its possible employment in perhaps from 10,000 to 20,000 crystallization plates, is then able to sharpen and school his observation and keenness to an extraordinary degree by paying attention to the smallest details and nuances of the forms and shapes of the crystals. It is in this perhaps that the value of

41

this work lies rather than in an abstract drawing of conclusions from this or that "indication". The general dynamic relationship of the different groups of crystals alone offers already a stimulating study, for example, the conclusions concerning the vitality and constitution of the patient which can be drawn from the harmony or disharmony of the crystallization picture.

There are also technical difficulties which interfere with the possibility of a general employment of this method which are carefully described in the chapter on "Technical Details". Anyone who carefully reads this chapter and follows its directions in his laboratory practice, will soon notice that an exact work with this crystallization method places a high demand upon the precision of the experimenter. If he does not in the beginning attain immediately the results here stated, he can be sure that his carrying out of the experiments was faulty. Co-workers and students who have been active in our research laboratory for a number of years, all know how much patience and care is demanded of them in order to produce an exact crystallization plate. They all know also that after a certain period of training it *was possible* to achieve the results here described. The individuals who have been trained in our laboratory were in the position with these methods to obtain also similar crystallizations in other localities. The possibility of an extended application of this method is offered in the comparison of various blood crystallizations of the same patient with the blood taken at different times, for example, prior to, at the beginning, during and after the conclusion of a determined treatment. As a result of changes in the blood crystallization picture, we may draw conclusions about the constitution of the patient. Besides this the treatment may be observed and followed in its effects on the patient.

IV. The Effect of Remedies and its Help in Judging Crystallizations

The formative forces which may be studied in the crystallizations made by admixtures of plant juices or of extracts made from bodily organs, offer in every case a partial picture of what may be seen in its entirety in the human blood crystallization picture. A knowledge of these forms is always necessary in order to come to a quick conclusion about a specific "centre". Sometimes it is possible to discover in the processes of disease a healing process corresponding to it.

The following illustrations present merely a "collection" of the form material which can be "consulted" when considering a blood crystallization picture as has already occurred with the forms previously described. The principle may be described by means of two cases. As to the rest they are presented without comment.

Plate XXIV

Fig. 70. 5% copper chloride solution with admixture of 15 drops of a preparation D 2 made from the retina.° Here only the interpretation of the characteristic part of the form is offered:—The radiating centre is striking; some of the rays however are more vigourously accentuated than others; between these there are open spaces.

Fig. 71. 20% copper chloride solution, mixed with blood taken from a case of detachment of the retina. We find in the second quadrant (the head zone) a form, somewhat disturbed, nevertheless with the evident tendency toward a similar sort of radiating centre. The difference in the density of the two crystallizations 70/71 consists in the fact that we usually make the

---

° This preparation as well as all of the following preparations were kindly placed at our disposal for the purpose of experiment by the Weleda Co. Arlesheim, Switzerland. I am greatly indebted to Mr. Johannes Teichert, Berlin, for the carrying out of this part of the research.

crystallization of the organic preparations with a 5% to 10% solution of copper chloride, on the other hand we use a 20% solution in making the blood crystallizations. In judging and comparing the difference between the crystallizations made with preparations derived from "organs" and crystallizations made with blood this must be taken into consideration. In the 20% solutions the forms always appear somewhat denser and more compressed than in 5% solutions. We may also compare the forms of this crystallization picture with the forms characteristic of tuberculosis, for which they may be mistaken; from this we learn to observe how "delicate" is the variation of form.

Fig. 72. An enlargement of the typical form: admixture of 15 drops of a preparation of "pelvis renalis" D 3 with a 10% copper chloride solution.

Plate XXV

Fig. 73. A 5% solution of copper chloride with an admixture of 1 drop of belladonna, mother tincture. The bent, elongated cavity with its transverse ray is characteristic.

Fig. 74. A detail from the middle zone of a crystallization of human blood showing the "belladonna-form". This presents a case among others of a nervous, spasmotic cough. As for the rest only the typical forms are offered. They are a selection from over 70 different preparations.

Fig. 75. 10 cc. of 5% $CuCl_2$, admixture of 7 drops of a preparation of medulla spinalis D 2.

Fig. 76. 10 cc. of 5% $CuCl_2$, admixture of 5 drops of a preparation of medulla oblongata D 2.

Fig. 77. 10 cc. of 5% $CuCl_2$, admixture of 5 drops of a preparation of medulla oblongata D 2.

Fig. 78. 10 cc. of 5% $CuCl_2$, admixture of 7 drops of a preparation of corpus quadrigeminum.

In these cases we were interested in seeing how far it was possible also to study the activity of homoeopathic remedies. That the so called "mother tinctures" are active, is self evident, because they are pure extracts. For every individual case of this sort, the most effective dilution and number of drops must first be discovered through untiring empirical laboratory work. The results were so interesting that the research has been carried further and will perhaps be published later on.

In the present plates from 70 to 78 it is not yet a question of asking whether it is homoeopathy or not, since the given dilution lies *clearly* in the region of the measurable. Partial influences on the forms of the crystalliza-

44

tion picture can be observed in the case of the so called "homoeopathic high potencies"; these experiments however place great demands on the exactness of the experimenter. We would not suggest making such experiments in order to become acquainted with our method.

These crystallization pictures are valuable in pointing out the direction in which an observer must seek in analyzing a "case".

V. Technical Details

The crystallization method described in this book is designated "sensitive". The crystallization processes react not only to the fine formative forces of the admixed substances under investigation, but also to all the most minute mechanical and atmospheric influences and disturbances. In order to avoid these extrinsic influences, and to have as a result in the finished crystallization picture only the changes caused by the formative forces of the admixture, it is necessary to make all preparations with the greatest of care.

Foremost among the difficulties of carrying out this crystallization method in other laboratories is a disregard, by those who attempt it, of the instructions given in this brochure.

CRYSTALLIZATION CHAMBER. This should be vibration-free with controlled temperature and humidity. The chamber is best built in a quiet room of constant temperature, ideally used for no other purpose. The solid floor is best of concrete or steel construction and in such a location that there is no vibration from surrounding rooms or corridors. Likewise the crystallization chamber should not lie within an electric or magnetic field.

Since it is preferable to crystallize only one to two cases at a time in one chamber, it is necessary to have more chambers if one intends to handle more cases a day. It is ideal to have one general room containing only the crystallization chambers. One ventilator, preheater and predrier can thus be used for all of the cubicles. Another room would be needed for the preparatory work.

In order to test for vibration, fill a Petri dish with water and place on the proposed surface. Then pour a small quantity of seeds of the Lycopodium plant onto the water so as to form a thin film. If the light reflection shakes visibly, this indicates that there is perceptible vibration.

Preferably the chamber should not be near a cold hall nor another cold room, or near a window. If the room where the chambers are to be placed has an outer wall, it is best to leave an air space between the outer wall and the wooden wall of the cubicle. A source of fresh pure air should be available.

CONSTRUCTION. The dimensions of the chamber should be approximately as follows: width—65 inches; height—85 inches; depth—65 inches.

In order to be of constant temperature over the 14-19 hour periods needed for the crystallization, the cubicle must be air tight and protected against loss of warmth by an insulated double wall. Wooden frames (made up in the prescribed dimensions of 2 x 2's) are covered on both sides with sheets of plywood, ¼ inch thick.

The absorption of impurities from the air in the room must also be avoided. Therefore all the walls should receive a coating of Cohesan, paraffin or shellac so that they are washable and not absorbent. If paraffin is used it is spread with a small hot iron. All joints must also be made air tight with paraffin. A wax spray (aerosol) can be used too, as well as any of the modern synthetic plastics, such as polyethylene. But make sure that the material is not volatile, does not give up fumes, and that the solvent has been completely removed from the chambers by thorough ventilation.

The *door* of the chamber should be insulated with strips of rubber tubing in the joints. There is an observation window in the door approximately 10 by 10 inches. The door is closed with two locks or latches.

TEMPERATURE automatically controlled is best between 28 and 32 degrees centigrade with the average of 30 degrees. With every change in the outside temperature and therefore a difference in draft direction (due to a different rate in the cooling of the outer wall of the chamber) we will be likely to get an error. Consequently it is best to have the surrounding laboratory as constant in temperature as possible. We now use a plywood enclosed space heater which produces a very low heat, so that the heat flow is more steady and less turbulence of the air is produced by sudden, strong heat waves. Lignotherm Heizplatten, 500 x 500 mm., 220 Volt, 30 watts Blomberger Holzindustrie B. Hausman G.m.b.H., obtained from Blomberg, Lippe, Germany. For each chamber, 4 plates are suspended underneath the ceiling; 2 plates are placed on the floor.

HUMIDITY must be kept between 35 and 55% with the average around 45%. It is measured by a Humidiguide or wet and dry bulb thermometer. Raising the temperature dries the air. Therefore, the first means of drying air is by raising the temperature up to the limit of 30 degrees centigrade. If this is not effective, an artificial drying will be necessary. This can be done either by drying the ventilation air in advance (if the air is very moist) or by placing 1, 2, or 3 Petri dishes of absorbent Silica Gel in the crystallization chamber, as much as is necessary to bring the humidity down to at

47

least 50%. The used Silica Gel can be renewed every day by heating above the boiling point. However, sudden changes of barometric pressure and weather can influence the evaporation time, and the crystallization in consequence.

It has been observed that special provision must be made for difficulty in the control of humidity during summer. If the outside temperature is as high or higher than the prescribed crystallization temperature of 30 degrees centigrade, it is impossible to dry the air by raising the temperature. To accomplish this it is necessary to have a difference in temperature of 5 to 6 degrees. A result of too wet air in summer (above 60% humidity) is slow evaporation time, e.g. more than 20 hours. This causes an enlarged crystal formation, which in general is less reliable for diagnostic purposes than that described in this paper. If therefore, enlarged forms are constantly obtained, the conditions of crystallization will have to be changed.

Under opposite conditions, when the air is too dry, the temperature can be lowered to the limit of 28 degrees centigrade or, especially in winter, a few Petri dishes of water (for evaporation) can be put in the chamber.

It is very important to regulate the temperature and humidity within the given limits, otherwise the results are less reliable. Too dry air shortens the crystallization time and produces rather one-centered atypical pictures. Too wet air results in too long a crystallization time and produces too many disturbances and an irregular pattern. The correct time of evaporation is more than 14 and less than 20 hours.

VENTILATION. Very careful ventilation between each series of tests with fresh air filtered to remove all fumes, dust and impurities must be carried out. A ventilator blows fresh, filtered air into the crystallization chamber through pipes of at least 1 inch in diameter. The size of the ventilator should allow renewal of the air in a chamber 3 to 4 times within one to two hours, i.e. it should have a capacity of 300 to 500 cubic feet per hour.

The renewal of air takes place only between two crystallizations, when the chamber is not in use. The used air should leave the room via another duct which leads outside the building so that it does not interfere with the rooms and the intake of fresh air. During crystallization the two ducts should be tightly closed in order to avoid disturbing drafts in the chamber.

TABLE. The chamber is furnished with a glass table in an iron frame suspended on a small link chain provided with microscrews for leveling the table. The chain acts as a shock absorber. The table is weighed down with a heavy lead weight. Two tables are suspended in each chamber which makes it possible to have two crystallizations done simultaneously. It is not necessary to wash and clean the entire surfaces of the chambers every time or every day. Only the tables should be cleaned and wiped with a non-dusty cloth or silica-treated paper. The walls and floors may be washed and cleaned every two weeks or, if infrequently used, once a month.

PREPARATION OF THE GLASS PLATES. The glass on which the crystallization is allowed to occur must be of good quality without scratches, air bubbles, rolling stripes, or rippled. The best quality of window glass, single thickness, is sufficient. Glass unfortunately, is very irregular. Unevenness can be discovered easily. If the glass is not even, glass errors will be recognized by the outcome of the crystallization. Plates with glass errors must be discarded. The plates are either round 4″ in diameter or 4″ x 4″ squares.

Wash the plates first in a hot 2% $NaCO_3$ solution, renewing the solution frequently if plates are very dirty. Wipe off with sterile cotton, using a new wad after about every tenth plate. Rinse in cold water. Rinse in 1% HCl solution. Rinse thoroughly in distilled water. Dry with a soft, non-fuzzy towel or cleansing paper. Keep wrapped in filter paper. Plates should not scratch each other or glue together. Work as dust-free as you can and keep the plates protected against dust. Each particle of dust on the plate may act as a seeding-germ. Any grease or fingerprint will impair the perfect surface which is necessary for a successful crystallization. Handle plates only at the edge; never touch the surface. Washing, etc., is best done with rubber gloves so as to avoid sweat and lint from the hands contaminating the plates. Keep warm (30°C) for use. The plates should be at this temperature for putting on the cellophane rings.

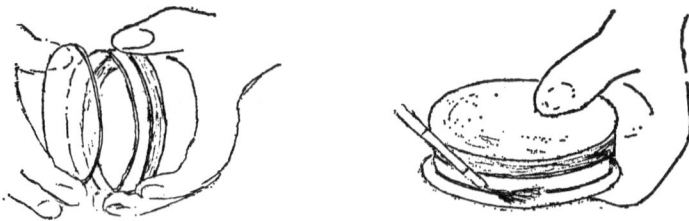

Drawing 4

Cellophane rings are then fixed to the plates so as to form a shallow, circular dish. These cellophane strips are 1 millimeter in thickness, 1 centimeter wide and 31 centimeters in length.

1. The glass plate is held against an aluminum dish around which one of the cellophane strips has been placed, with its ends paraffined together. The outer edges of the strip and the glass plate are then sealed with paraffin,

applied with a small camel's hair brush, care being taken that the cellophane strip and the plate touch at all points. Once the paraffin wax has set the aluminum dish is removed. (See sketch.)

2. The cellophane ring must not be touched inside and is carefully and equally pressed against the glass plate to prevent the paraffin from leaking in under the ring. The paraffin used is mixed with a small amount of opal wax and has a melting point of about 160 degrees Fahrenheit.

PREPARATION OF CUPRIC CHLORIDE SOLUTION: Kahlbaum's $CuCl_2 \cdot H_2O$ was used prior to 1940. Since 1950, we use $CuCl_2 \cdot H_2O$ manufactured by Merck and Co., Darmstadt, Germany. They produce a special brand for crystallization according to Pfeiffer. This is obtainable through Th. Geyer A.G., Stuttgart, W. Germany. No other $CuCl_2$ has been found suitable enough. The $CuCl_2 \cdot H_2O$ should consist of fine needles. Clusters, tablets, platelets and crusts of $CuCl_2 \cdot H_2O$ crystals have to be picked out with a porcelain spatula for they do not give the desired result; in fact, they cause disturbances belonging to another modification. $CuCl_2 \cdot H_2O$ is somewhat unstable. This can be overcome by keeping a small vial with HCl (1%) suspended in the bottle over the $CuCl_2$ so that the atmosphere is saturated with HCl fumes. Under no circumstance should the HCl come in contact with $CuCl_2$ crystals directly.

A 20% solution of $CuCl_2$ (freshly prepared) is used for human blood; a 15% solution for surgical specimens, animal and children's blood, a 5% solution for plant extracts.

PREPARATION OF SAMPLES: *Specimen extraction solution:* (other than blood)

1. The specimen to be extracted should be as fresh as possible, not over 6 hours old. It should not be put in alcohol or formaldehyde, but should be kept in a test tube with a cotton stopper and immersed in a water bath at 37° centigrade (blood temperature). Specimen can be kept frozen until used.

2. One cc. of distilled water and a very small piece of the specimen (about ½ cubic inch) are placed in a small mortar. The specimen is cut into small bits and pressed with a pestle. Four cc. of blood-warm distilled water are added and the mixture allowed to stand until the solid matter settles to the bottom.

3. To each of several test tubes (depending on the number desired for diagnosis) containing 8 c.c. of a 15% solution of Chloride of Copper at blood temperature, 0.5 c.c. of the extract is added by means of a pipette. The solutions are mixed, as in the case of the blood, and taken to the crystallization chamber.

SOLUTION FOR BLOOD DIAGNOSIS: 0.1 to 0.2 cc. of a 6% hemolized capillary blood solution are slowly introduced with a pipette, into 10 cc. of a 20% solution of Chloride of Copper. This is carefully mixed by slow rotation of the test tube 20 times (not by shaking) and poured on the glass plates in the crystallization room. Not less than five plates should be taken for the diagnosis of one case. The blood solution should not be older than 30 minutes, and there should never be a coagulated clot or fiber in it when it is mixed with the Chloride of Copper and poured out. Although the blood solution should not stand longer than 30 minutes, the blood itself can be preserved and shipped in the following manner. The capillary blood is dropped directly from the finger tip on a clean disk of filter paper. As soon as the blood has been absorbed by the filter paper and no longer runs on the surface, the disk is placed between two others and slipped into a cellophane envelope. This can then be sent in a letter by air mail. It has been found that the blood is effective for crystallization up to 5 days after the sample has been taken. The sample is removed from the disk by soaking the impregnated paper in a small amount of distilled water for a short time. Then the usual dilutions and additions are made as described above. We owe this technique of preserving the blood samples to a painstaking study by Dr. Frieda Bessenich of Dornach, Switzerland.

STEPS IN ACTUAL TEST:

1. Control of temperature and humidity. The morning after the crystallization is completed the room should be cleaned once the plates have been removed. Stop the heating and start the ventilation.

2. About noon, that is, 3 hours after the first step, stop ventilation and start heating.

3. Immediately before placing the new series in the chamber, check temperature, humidity and make necessary corrections. The chamber should have reached the correct temperature before the new series is put in.

4. The ventilation has been turned off and the ducts closed. The plates are leveled with a small capenter's level (4 inches in length). The plates must not be too near the wall, so as to avoid drafts. It may be that some positions are less successful than others, due to air currents, etc. The best places may be determined by checking every square 4 inches and observing the regularity of forms. Number the squares and report which give better and which poorer results.

5. The plates should be put in the room just before crystallization, after the leveling is completed and everything is ready for the solution to be poured into them.

6. In pouring out the solution, care must be taken to see that the entire bottom of the plate is immediately completely covered over evenly, other-

wise the picture will show some disturbance in forms. A solution poured too quickly, with a great sweep, will show featherlike forms.

7. The chamber is left at once and the door quietly shut and the crystallization allowed to proceed without disturbance for 14–19 hours.

8. After removal from the crystallization room, the cellophane strips are removed, rinsed, scraped and washed with hot (but not boiling) soap and water, rinsed again, rubbed with alcohol and placed in a Petri dish to keep them in form until used again.

9. The plate, after the paraffin is removed, may be a) preserved by sealing another plate on its surface with waterproof adhesive, b) covered with plastic or c) photographed, and the negative alone kept.

PHOTOGRAPHIC TECHNIQUE: We use a Leica III 35 mm. with an Elmar fl 5 cm. 1:3.5 lens. The film is Kodak High Contrast M. 35 mm, 417 100 ft., extreme resolution Panchromatic safety film, from which we cut suitable lengths to load the cartridges. Developed in a daylight tank Rondinax 35 U (made in Germany) with Promicol ultrafine grain developed from May Baker, Ltd., Dagenham, England.

The exposure is ¼ sec. at f= 14 to 16.

This way we get the finest possible resolution. The finished, dry negatives are cut and placed in plastic jackets, which hold 6 exposures; size of jacket 5x3¼ inches. For each case, one jacket is used. The negatives can be inspected without removing from the jacket, either with a projector projected on a screen or a viewer, enlarging it to natural size, a so-called "projector table viewer".

The importance of following the specific technical details can also be recognized if one observes the process of the crystallization itself in its single steps. This has been done in our laboratory with the help of a movie camera. The resulting pictures taken during the whole time of the crystallization process (about 15–19 hours) show definite phases which can be described shortly as follows.

PHASE A. During this first phase, lasting 10-14 hours, the solvent (water) evaporates slowly, the $CuCl_2$ solution becomes concentrated. There are no visible changes in the liquid.

PHASE B. Slight turbidities appear. These turbidities, caused by partial settling out of colloidal protein matter, show already localization and form elements. It is a kind of preformation of the final picture. Therefore, this phase (1½–2 hours) is the most critical one and should under no circumstances be disturbed.

PHASE C_1. Also this is a very critical phase. During this phase (30–40 minutes) the actual crystallization of the typical forms begins. The solvent has already so far evaporated that the $CuCl_2$ solution is somewhat over-saturated and crystallization is possible. It may start on one point but also on several points at the same time, showing the main forms as quite independent one from the other.

PHASE C₂. In this next phase the empty places between the main forms are filled in without any change of the already manifest main forms. All single forms are now combined into the final crystallization picture. This phase lasts about 90 minutes.

PHASE D. In this end phase the outer edge zone is formed and the last remainder of the solvent evaporates. This phase is the least sensitive one to disturbances.

It should be emphasized that the duration of the single phases as mentioned above is valid only for the blood crystallization process under the above prescribed technic. They will be different when other concentrations, other additions, etc. are used.*

DETECTION OF TECHNICAL ERRORS:

Cleaning errors produce incrustations, a thick heap of crystals, or spaces not covered at all.

Fingerprints can easily be detected. The crystals on them are less thick, criss-cross and not coordinated with the surrounding plane. They are most apt to occur near the edge zone. The finger form is easily distinguished.

Paraffin underneath the rim, penetrating towards the center of the plate causes a circle of little crystallization centers all around it or a long radiating form oriented away from the periphery.

Grease on the cellophane rings creates much disturbance on and near the edge zone and many little irregular crystals.

Leveling errors or ripples in the plate often result in the crystallization being divided into two parts, or the gravity point is near one side, the other part showing a large edge zone with a thick end in the middle.

Wind errors are very apparent, the crystals having a wind-blown appearance.

INTERPRETATION: Plant extracts are variable according to the seasons of the year and the crystallizations themselves show that it is difficult to obtain a clear picture during July and August, while during the winter months from October to May satisfactory results are obtained. Furthermore local constants play a roll in the crystallization formation. It has been noticed for instance, that under otherwise similar conditions the same substance in Egypt has the tendency during crystallization to form more deli-

---

* For further details see the article of Dr. E. Pfeiffer in the book: "Zur Methodik der Empfindlichen Crystallisation" by F. Bessenich.

53

cate and satin-like crystals. The Middle European type is reproduced here in this brochure. In Holland one has to contend with excessive atmospheric humidity. In North America crystallization pictures were produced in which the single individual crystal groups or rays were more densely pressed together, forming a more compact mass. These local influences must also be taken into consideration. They are to be found again in the whole of nature just as soon as formative forces are able to be effective. One and the same mineral substance often shows variations in crystallization in different parts of the globe. Nevertheless by following the instructions given in this work it has been possible to judge a copper chloride crystallization made in Dornach just as successfully as one made in Strassburg, Munich or New York. This we would like to emphasize especially.

Questions are often asked in regard to "proofs". If someone makes a chemical analysis he is generally satisfied if an analysis agrees to the decimal point or if in the case of a total chemical analysis a 100% result is attained with 0.5% variation or tolerance. In the experiments with living material there is a certain zone of error which can be narrowed by repetition and different methods of calculation. In the technic of making crystallizations the exact similarity of the pictures is about 80%. This of course only holds good for the experienced researcher, not for the preliminary experiments of the novice.

In the description of the type of pictures, somewhat different rules should be considered. Every picture type is absolutely unique. This is connected entirely with the nature of the structure per se. It has to do here more with physiognomical laws. A physiognomical symbol is determined by means of itself. A rose is a rose and remains a rose, whether it is determined by one example of its species or by 500 examples.*

The description of a form-type in nature remains therefore also clear whether it is determined in one case or in several. Naturally there have

---

* Reference is here made to the exposition of Rudolf Steiner in the preface to Goethe's Natural Scientific Works (published by the Union Deutsche Verlagsgesellschaft, Stuttgart vol. 1 p. LXI.) and Goethe's own development of the theme "perceptive power of thought" and "preliminary sketch of a general introduction to comparative anatomy."

Rudolf Steiner p. LXI. "The idea through which we apprehend the organic is therefore essentially different from the concept by which we explain the inorganic. The former (idea) does not merely collect a multiplicity of facts like a sum, but produces its own content from itself. It is the result of the given (experience), the concrete phenomenon. This is the reason why we speak of laws in inorganic natural science—natural laws—and explain facts by them; but in organic nature we do this by describing the form types. The law is not identical with the manifoldness of the conceptual world which it rules. The law is superior to the manifoldness of the concept. In the type however the real and ideal become a unity; the manifold can only be explained as proceeding from a point of the whole, i.e. from a point of a manifold nature identical with the whole. In the knowledge of this relationship between the science of the inorganic and of the organic lies the significance of Goethe's research."

been so many repetitions with similar results, giving a high degree of certainty, that everyone is able to discover and to recognize the true form-type and not to mistake a rose for a lily.

Hence in the description of a disease-form, a few instances will suffice in order to establish the exact facts of a case. The English physician, Th. Addison, described in 1855 the disease named after him** so well and exhaustively, that all the subsequent research of others has added only very little that is essentially new. We would even go so far as to say that anyone who is not able to define a form-type from a few cases will not be able to accomplish it from a hundred examples. Naturally it is not intended to underrate the value of repetition in order to gain greater accuracy. Even now while these lines are being printed the range of experimental material has been enlarged. Every one of these very latest cases would be suitable for quotation here. This is mentioned in order to accentuate how continuous work with this method increases its certainty. It is not the intention here to publish something sensational, but to report the facts of years of earnest labor.

** R. G. Hoskins: "The Hormones and the Life of the Body" p. 12. "He has seen 11 cases. With the exception of omitting some nervous symptoms this description has not been surpassed."

VI. BIBLIOGRAPHY

Alberti, G.—La diagnosi microscopia. Reazione cristallina di Pfeiffer. L'Artulita Medica, IV, No. 5-6; 152.

Beckman, H.—Ueber Keimbildung, Eiskristallwachstum und Aufflaecherungswachstum von $CuCl_2$. H_2O in rein waessriger und Eiweiss-haltigen Loesungen. Inaug. Dissert. am Mineralog. Institut der Universitaet Bonn. (1959). (About germ formation, icecrystal growth and the spreading of $CuCl_2$ crystallization growth in pure water and protein containing solutions.)

Bedding, W. C.—Merkwaardige kristallisatieverschynselen veroorzakt door de aanwezigheid van biymengselen. 7th Dutch India Science Congress, Batavia, 1935.

Begouin, M. P.—Quelques résultats de la méthode des cristallisations de Pfeiffer dans le diagnostic du cancer et de la tuberculosis. Bulletin de l'Académie de Medecine. Paris. Tome 119, No. 25, p. 746. (Some results with the Pfeiffer crystallization method in the diagnosis of cancer and tuberculosis.)

Bessenich, F.—Goethe in unserer Zeit. (Goethe in our time). Containing results obtained with plant extracts. Editor: Natural Science Section, Goetheanum, Dornach, Switzerland. 1949.

Bessenich, F.—Der Kristallisation Test. (The Crystallization Test.) Weleda Nachrichten, No. 66. 1954.

Bessenich, F.—Zur Methode der Empfindlichen Kristallisation. Naturwissenschaftliche Sektion am Goetheanum, Dornach 1960. (About the method of Sensitive Crystallization.)

Bourgois—Diagnostic Precoce et Traitement du Cancer. Paris 1954. (Early diagnosis and treatment of cancer.)

Engquist, M.—Gestaltkraefte des Lebendigen. Die Kupferchlorid—Kristallisation eine Methode zur Erfassung biologischer Veraenderungen pflanzlicher Substanzen. Vittorio Klosterman. Frankfurt am Main. 1970. (Formative forces of living substances. Copper chloride crystalization, a method for demonstrating biological changes of plant substances.)

Engqvist, M.—Struktur-Veraenderungen im Kupferchlorid-Kristallisationsbild von Pflanzensubstanzen durch Alterung und Duengung. (Structural changes in the copperchloride crystallization picture caused by plant substances through aging and fertilization.) Lebendige Erde, Heft 3. 1961.

Friess, H.—Ueber eine neue kristallographische Methode zum Nachweis von Veraenderungen in konservierten pflanzlichen Lebensmitteln. Mitteilung aus der Reichsforschungsanstalt fuer Lebensmittelfrischhaltung in Karlsruhe. 1944. Zeitschrift fuer Lebensmittel-Untersuchung und Forschung. Band 87, Heft 1/3. (A new crystallographic method to demonstrate changes in preserved vegetable foodstuffs.)

Garn, Wm.—Blutkristallreaktion auf pathologische Stoffwechselprodukte beim Karzinom. Archiv fuer Geschwulstforschung, Band 2, No. 3. Verlag Steinkopf, Dresden 1950. (Blood crystallization Reaction to Pathological Products of the Metabolism in Carcinoma.)

Gruner, O. C.—Experiences with the Pfeiffer Crystallization Method for diagnosis of cancer. Journal of the Canadian Medical Association 43; 99-106. August 1940.

Gruner, O. C.—A study of the Blood in Cancer. Renouf Publishing Co., Montreal, Canada. 1942.

Hahn, F. V.—Thesiegraphie. Untersuchungsmethode an Biologischen Objekten, insbesondere Nahrungsmitteln. Franz Steiner Verlag. Wiesbaden 1962. (Thesiegraphy. Research method on biological objects, especially food.)

Hollemann.—Ein Beitrag zum Verstaendnis der Empfindlichen Kristallisation. . Elemente der Naturw., 4, S.24. 1966. Philos. Anthrop. Verlag am Goetheanum, Dornach, Schweiz. (A contribution for the understanding of the Sensitive Crystallization.)

Isabel, E.—La Cristallisation du Chlorure de Cuivre. Son application au diagnostic du cancer et de la tuberculose. . 1940. Jouvre & Cie. (The copperchloride crystallization. Its application in the diagnosis of cancer and tuberculosis.)

Jung, H.—Beitraege zur Kristallographischen Blutuntersuchung. Pharmascia 7, H.10; 628-239, 1952. (Contributions to crystallographic blood tests.)

Koepf, H. & Selawry, A.—Application of the Diagnostic Crystallization Method for the Investigation of Quality of Food and Fodders. Bio-Dynamics No. 64 and 65. 1962 and 1963.

Koopmans, A.—Zeitabhaengigkeiten bei "Empfindlichen Kristallisationen". February 1965. (The time element in Sensitive Crystallizations.)

Kraus, H.—Copperchloride crystallization test for the detection of aging and decomposition of meat, fish and milk. Arch. Lebensmittelhyg. 10, 35-40. 1959.

Krebs, H.—Ein Beitrag zur kristallographischen Carcinomdiagnostik nach E. E. Pfeiffer. Helvetica Chirurgica Acta. Vol. 14, 1947, Fasc. 3. (A contribution to the crystallographic diagnosis of cancer according to E. Pfeiffer.)

Krueger, H.—Beitrag zum Studium der Bildekraefte von Organen und Heilpflanzen. (Contributions to the Study of the Formative Forces in Organs and Remedial Plants.) May 1949.

Krueger, H.—Kupferchlorid Kristallisationen, ein Reagenz auf Gestaltungskraefte des Lebendigen. Weleda Schriftenreihe. Hippocrates 1949. Stuttgart. (Chloride of Copper Crystallizations, a Reagent upon Formative Forces in the Living.)

Krueger, H.—Tageszeitenrhytmen bei Pflanzen. Weleda Nachrichten, 22 (1949) Stuttgart. (Daily Rhythms of Plants.)

Kubina, H.—Die Blutkristallisation nach Pfeiffer. Wissenschaftl. Haus Mitteilungen 24 (1945) 8, 221. (The Blood Crystallization according to Pfeiffer.)

Lehmann, F. Grube—Ueber die Beeinflussung der Kupferchlorid Kristallisation durch liquor cerebrospinalis. Dissertation. Hamburg, 1953. Also in: Arch. Psychatrie und Zeitschrift fuer Neurologie 192 (1954) 207-219. (About Influencing the $CuCl_2$ Crystallizations by means of Cerebrospinal Fluid.)

Mackaye, M. C. J.—-Onderzoekinge over gevoelige Kristallisaties volgens E. Pfeiffer. Inst. voor praevent. geneeskunde, Amsterdam, 1941, 36-48. (Research with Sensitive Cristallizations according to E. Pfeiffer.)

Merten, D., Lagoni, H.—Die Kuperchlorid-Kristallisation als analytisches Hilfsmittel in der Milchforschung. Milchwissenschaft 13, April 1958, p. 159-161. (The copperchloride crystallization as an analytical tool in dairy research.)

Merten, D., Lagoni, H., Peters, J. H.—Ueber den Einfluss von Milch und Milchbestandteilen sowie Milchprodukten auf das Kupferchlorid-Kristallisationsbild. Kieler Milchwirtschaftliche Forschungsberichte XI, 1, p. 69-79, 1959. (About the influence of milk and milk constituents as well as milk products upon the copperchloride crystallization picture.)

Merten, D., Lagoni, H., Peters, K. H.—Ueber Bestandteile der Milch als Loesungsgenossen bei der Kupferchlorid-Kristallisation. Proceedings of XV International Dairy Congress 1959, 3, 1739-42. (About constituents of milk as factors in solution in the copperchloride crystallization.)

Miley, G.—Crystallography. The Hahnemann Monthly, May 1940. Philadelphia, Pa.

Morris, D. L., Morris, C. T.—Specific Effects of certain Tissue Extracts on the Crystallization Pattern of Cupric Chloride. Journal of Physical Chemistry, Vol. 43, No. 5, May 1939.

Morris, D. L., Morris, C. T.— The Modification of Cupric Chloride Crystallization Patterns by Traces of Proteins. Journal of Biological Chemistry, Vol. 43, No. 5, May 1939.

Neuhaus, A.—Was sagt die Wissenschaft zur Kupferchlorid-Diagnostik? Umschau, Heft 6, 1960. (What does science say about the diagnostic copperchloride method?)

Nickel, E.—Die Reproduzierbarkeit der sogenannten "Empfindlichen Kupferchloridkristallisation". Mineralog. Institut der Universitaet Freiburg, Schweiz. Bull. Soc. Frib. Sc. Nat., Vol. 57, Fasc. II. 1967/68. (The possibility of reproducing the so-called "Sensitive copperchloride crystallization".)

Petterson, B. D.—Beitraege zur Entwicklung der Kristallisations Methode mit $CuCl_2$ nach Pfeiffer. June 1967. (Contributions to the development of the crystallization method according to Pfeiffer.)

Pfeiffer, E. E.—Kristalle. Orient-Occident Verlag, Stuttgart, Den Haag, London 1930.

Pfeiffer, E. E.—Studium von Formkraeften an Kristallisationen. Naturw. Sektion, Goetheanum, Dornach, Switzerland. 1931. (Study of Formative Forces in Crystallizations.)

Pfeiffer, E. E., Knauer, I.—La Cristallisation Sensible. L'Homeopathie Francaise. 1933. Paris. (The Sensitive Crystallization.)

Pfeiffer, E. E.—Les forces éthériques en biologie humaine et végétale. Bulletin du Centre Homoeopathique de France. No. 1. 1934. Paris.

Pfeiffer, E. E.—Empfindliche Kristallisationsvorgaenge als Nachweis von Formungskraeften im Blut. Verlag Emil Weise, Dresden. 1935.

Pfeiffer, E. E.—Sensitive Crystallization Processes as Demonstration of Formative Forces in the Blood. Verlag Emil Weise. Dresden. 1936.

Pfeiffer, E. E.—Revue generale de la science pure et appliques. No. 47, 424. 1936.

Pfeiffer, E. E.—Formative Forces in Crystallization. Anthroposophic Press, New York. 1936.

Pfeiffer, E. E.—Le diagnostic des maladies végétales, animales et humaines par la méthode des crystallisations sanguines. L'Homeopathie Française. Paris. May, 1938.

Pfeiffer, E. E.—Ueber die Beeinflussung des Kristallisationsbildes des Kupferchlorides durch tuberkuloses Material. Muenchner Medizinische Wochenschrift. No. 3, 92. 1938.

Pfeiffer, E. E.—The Influence of Blood of Malignant and Nonmalignant Origin upon the Crystallization of Copper Chloride. Paper read before the Third International Congress on Cancer, Atlantic City, New Jersey, Sept. 1939.

Pfeiffer, E. E.—Sensitive Crystallization. Chemical Products & The Chemical News. Vol. 3, No. 3. London. Jan. 1940.

Philipsborn, H. V.—Biomineralogie. Fortschritte der Mineralogie, 32 (1953) 11-25.

Riebeling, C.—Ueber die Beeinflussung der Kupferchloridkristallisation durch Koerpersaefte. Photographie und Forschung. 1954. No. 1. (About the Influence of Body Fluids upon the $CuCl_2$ Crystallization.)

Selawry, Alla—Die Pfeiffersche Blutkristallisations Methode. Wuerttembergisches Aerzteblatt, 1. Jahrgang, Heft 6. 1946. (The Blood Crystallization Method of Pfeiffer.)

Selawry, Alla—Neue Ergebnisse auf dem Gebiet der $CuCl_2$ Blutkristallisations Diagnostik. Deutsche Medizinische Wochenschrift. Stuttgart. Vol. 74, No. 8. 1949. (New Results in the Field of $CuCl_2$ Blood Crystallization.)

Selawry, Alla—Nachweis von Lungen- und Nierekrankheiten an Kupferchlorid-Blutkristallisationen. Stuttgart. Hippokrates, 25, No. 16, 1954, 510-514. (Demonstration of lung and kidney disease with $CuCl_2$ blood crystallization.)

Selawry, A.—Sind Herz- und Leber Krankheiten an $CuCl_2$ Blutkristallisationen nachweisbar? Munich-Berlin Medizinische Klinik 49, No. 9, 1954, 327-330. (Are heart and liver diseases demonstrable with the $CuCl_2$ blood crystallization?)

Selawry, A.—Possono accertarsi inflammazioni e tumori col metodo emo cristallografico al $CuCl_2$? Roma. Athena No. 1. 1954.

Selawry, A.—Kupferchlorid-Kristallisation als Reagenz auf Zustandsaenderung belebter Organismen. Hippokrates 31, 14, 31 1960. p. 453-460. (Copperchloride crystallization as reagent for changes in the state of living organisms.)

Selawry, A. and O.—Die Kupferchlorid Kristallisation in Naturwissenschaft und Medizin. Gustav Fischer Verlag, Stuttgart, 1957. (The copperchloride crystallization in natural sciences and in medicine.)

Selawry, O.—Ein Fall von Endocarditis. (A case of Endocarditis.) Acta Helvetica.

Selawry, O.— Ueber Querlagerungen im Blutkristallisationsbild gesunder und tumorkranker Menschen. Zeitschrift fuer Krebsforschung. 1955. (About oblique forms in the blood crystallization pattern of healthy and tumor-diseased human beings.)

Teichmann, B.—On the use of the Crystallization Test in Cancer. Arch. Geschwulstforsch. 25, 195. 1965.

Trumpp, J.—Nachpruefung der E. Pfeifferschen Angaben ueber die Moeglichkeit einer kristallographischen Diagnostik. Versuch einer Hormonskopie und Schwangerschaftsdiagnose. Muenchner Mediz. Wochenschrift 1939, No. 14, p. 544. (A check of E. Pfeiffer's indications about the possibility of a crystallographic diagnosis. Experimentation with hormones and pregnancy diagnosis.)

Trumpp, J. and Rascher—Versuch einer Kristallographischen Karzinom-Diagnose. Muenchner Medizin. Wochenschrift 1939, No. 14. (Experiment with a Crystallographic Diagnosis of Carcinoma.)

Toussaint, A.—Report on the use of the Crystallization Method for Control of the Quality of Milk. Die Taetigkeit der Stadt Strasbourg auf dem Gebiete der oeffentlichen Gesundheitspflege. Paris, Strasbourg. Societe Francaise d'editions d'art. Page 6 p. (The Activity of the City of Strasbourg in the Field of Public Sanitation.)

Von Hahn, F. V.—Thesiegraphie, eine neue Methode zur Qualitaetsbestimmung von Nahrungsmitteln. Hippokrates, 31. Jahrgang, Heft 7, 15. 1960. Hippokrates Verlag, Stuttgart.

Winsberg—Laboratoriumsmethoden zum Nachweis der Geschwulsterkrankungen. Stuttgart. Hippokrates, 1954. 26. (Laboratory Methods for Detection of Tumor Diseases.)

www.ingramcontent.com/pod-product-compliance
Lightning Source LLC
Chambersburg PA
CBHW051224200326
41519CB00025B/7241